T0130816

NOBEL CONFERENCE® XXV

THE END OF SCIENCE?

Attack and Defense

Sheldon Lee Glashow, Harvard University Nobel Laureate
Sandra Harding, University of Delaware
Ian Hacking, University of Toronto
Mary Hesse, Cambridge University
Gerald Holten, Harvard University
Gunther S. Stent, University of California-Berkeley

Edited by
Richard Q. Elvee

Gustavus Adolphus College
Saint Peter, Minnesota 56082

Copyright © 1992 by
Gustavus Adolphus College

University Press of America®, Inc.

4720 Boston Way
Lanham, Maryland 20706

3 Henrietta Street
London WC2E 8LU England

Library of Congress Cataloging-in-Publication Data

Nobel Conference (25th : 1989 : Gustavus Adolphus College)
The End of Science? : Attack and Defense / Nobel Conference XXV;
edited by Richard Q. Elvee.
p.   cm.
Includes bibliographical references.
1. Science—Philosophy—Congresses. 2. Science—Social
aspects—Congresses. I. Elvee, Richard Q. (Richard Quentin)
II. Title.
Q174.N6   1989   501—dc20   91-36080 CIP

ISBN 0-8191-8489-6 (alk. paper)
ISBN 0-8191-8490-X (pbk. : alk. paper)

 ^{TM} The paper used in this publication meets the minimum requirements of
American National Standard for Information Sciences—Permanence
of Paper for Printed Library Materials, ANSI Z39.48–1984.

This book is dedicated to our organized colleague
Elaine Brostrom, Director of Public Affairs,
Gustavus Adolphus College, 1978-1992,
who each year, brought it altogether, and made it work.

Table of Contents

Foreword

by William Dean, Department of Religion, Gustavus Adolphus College

The title, <u>The End of Science</u>?, asks not whether science itself is about to end or even to wane, but whether people will stop claiming that science knows nature as it is. Science, it suggests, may know nature only as the scientist sees it. Or the title suggests that, in knowing nature, scientists to some extent create nature.

No one bothers to ask philosophers or theologians, poets or politicians, workers or bosses whether they know the world as it is. It is common knowledge that the world for which they speak has been affected already by their description of that world. Will not the same fate strike scientists now? Has it not already? The person of the observer, in short, might affect always and ineradicably the meaning of the observed, even in science. If knowing something as it is can be called knowing something objectively, then the title questions the objectivity of the sciences.

This line of analysis would suggest that the meaning of our title itself must be challenged. The title, <u>The End of Science</u>?, might not be known as it is, in all its objectivity, but only as it is read to be. (This question about the meaning of observation might itself be a function of observation.) We should consider two types of readers and how they read the title: those who claim that the sciences are objective and those who claim that such claims now are dead.

Those defending the objectivity of the sciences will see the question as unjust. Must scientists be dragged, they ask, before the bar and accused of crimes committed by another people in another country? Was it not science that settled just that territory the humanities shunned when they chose Descartes' realm, where certain knowledge alone was worth having? Was it not science that chose the humbler world of probability, and worked modestly with mere hypotheses that were then tested and revised? Why then accuse scientists of mistakes committed by those obsessed with certainty and error-avoidance, when scientists were willing to settle for probability, to be critical about their realism, and to risk error if in the process new landmarks could be found? Must everyone automatically be chastised simply because some legitimately have been chastised? In their humbler, probabilistic world the scientists are able to know what they know, to corner the thing as it is--to be, after all, in some weak sense objective.

Those who deny the objectivity of the sciences will see this as an instance of special pleading. The title, <u>The End of Science</u>?, points to something no one can escape. If humanists are people and they know they cannot know the world as it is, uncolored by the coloration they inevitably lay on it, how is it possible for those other people who are scientists to know the world as it is? This question follows from just that audacious uniformitarianism--that generalizability--that the scientists themselves have protected even as they have worked in their humble world. Of course, when this question grows so powerful that even scientists are asking it, then we should expect to hear scientists' protests, their sense of having been misunderstood and wrongfully injured. Surely, the scientific <u>cri de coeur</u> will fill the air for decades, drown earlier ecclesiastical cries of wrongful reduction.

Those defending scientific objectivity will respond that humanists are pushing us all into a vortex of subjectivity, where no serious inquiry is possible. After all, they will argue, even humanists require objectivity. Do they not presume to be objective even as they claim that <u>always and everywhere</u> the sciences are not objective? When the critique of objectivity commits just the crime it protests, is this not another augury for objectivity?

Fleeing this labyrinth of language, we might better view the question, <u>The End of Science</u>?, on pragmatic grounds.

Those defending the objectivity of the sciences will see the question as dangerously distracting. Has it not been the freedom from just such epistemological recriminations that has given space for science's astounding success? Has not science by now done enough to guarantee that it grasps objectively some real and external world? Might not pressing the skeptical question be all too successful, throwing science into the same prison of words and theories in which the humanities have been confined for three centuries? Must humanists do to scientists what they have done to themselves? Should their jealousy and resentment be given license? Has it occurred to the new epistemologists, pluralists, relativists, historicists, and deconstructionists that they might actually be successful--luring the scientist out of the laboratory into twenty-first century coffee houses? Already there is a severe shortage of math and science students. Must this be aggravated by giving new venue to fashions of epistemological disillusionment? Is not more damage to be done to society if scientists are crippled irrecoverably with this hermeneutical fusillade? Disputes about the nature of knowledge and method may be the last arena where the humanist is still champion. Should the working scientist be thrown to the growling literati? If the desire for objective truth fuels the sciences, why not leave that desire undisturbed?

For the humanists it is simply too dangerous to placate the scientists by granting them honorary objectivity. Inevitably, the scientists will use their objectivist credentials to advance the subjective beliefs that all along attended their ostensibly objective knowledge. They may not realize that this is what they do (particularly if they socialize primarily with scientists) and neither do most of the rest of us, for the pretense of scientific objectivity hides and leaves unexamined the values that actually attend the sciences. Nevertheless, determinism, mechanism, positivism, and the worship of bigness all are worldviews and value-systems typical of scientists; and these views and systems have done their social damage in the nineteenth and twentieth centuries. In short, those who falsely claim objectivity inevitably smuggle their values into social decisions. Is this not suggested in environmental catastrophe, weapons that outrun the maturity of weapon-users, and health technologies that cause population explosions? Even if it is politicians and bureaucrats that legislate and administer such evils, does this not occur because they and their constituents already have traded their circumspection for a conscious or unconscious veneration of scientific objectivity and the values of scientists? And does not that veneration lead them to say that, if science-based technologies can be built, they should be built?

Or has the point of no return been passed, so that the only way safely to land is with more, not less, science? Is this the time to usher the linguistic dancing-masters into the cockpit?

Or is it just the time? Just in time?

Introduction

After Twenty Five Years: The End of Science!

One evening more than a quarter of a century ago, representatives from Gustavus Adolphus College and officials from the Nobel Foundation were sitting around at the country home of the Countess Bernadotte in Stockholm, Sweden, after the Nobel Prize ceremonies. The Gustavus constituency, which included then-President Edgar Carlson, Development Director Reynold Anderson and Philip Hench of Rochester (the only Nobel laureate in science from Minnesota to date), made a request. They asked the Nobel Foundation Board to endorse a series of science conferences at Gustavus, letting the Nobel name be used to establish credibility and high standards. The board granted the wish. And October 3 and 4, 1989, marked the 25th year of convening the world's top scientists and scholars for an intellectual festival.

"Genetics and the Future of Man," held in 1965 established the level at which the Nobel Conference would operate for the next twenty-five years. Four leading scholars in biology and ethics were joined by three Nobel laureates, Polykarp Kusch, William Shockley and Edward Tatum, in a discussion of biological engineering and genetic manipulations. With over 1000 people attending from 36 colleges and universities and 82 Minnesota high schools, Nobel Conference I, was at the very least a regional success. That success would expand rapidly. Nobel Conference III, "The Human Mind," attracted 2000 people to St. Peter in 1967 and was covered by the New York TIMES, the Los Angeles TIMES, and the Associated Press.

The attention lavished upon Nobel Conference III was due in large part to its topicality, as the participants discussed issues including mind control and psychedelic drugs. But this topicality pointed to one of the hallmarks of the Nobel Conference: This was not to be a dry recitation of arcane scientific facts, as the next several conferences amply demonstrated. The 1969 Nobel Conference on communications was highlighted by the speech of linguist and social critic Noam Chomsky. Noted African-American photographer, writer and filmmaker Gordon Parks illuminated 1970s Nobel Conference VI, "Creativity," which also featured a sharp debate between Jacob Bronowski and William Arrowsmith on the sources of creativity. A running debate over the two-day period was not unusual. In 1971, one of the most heated confrontations took place between Nobel laureate Glenn Seaborg and theologian Joseph Sittler, who argued the role of science and technology in aiding humanity (Seaborg) or in destroying the environment (Sittler). And in 1972, theologian Krister Stendahl and biologist George Wald both suggested that "the End of Life" was not succeeded by the immortality of the ego. Letters to editors poured into Minneapolis papers and religious journals protesting the comments of Stendahl and Wald. Not a few of those letters questioned the integrity of the college and its conference, questions that were likely not answered by Nobel Conferences IX and X--"The Destiny of Women" and "The Quest for Peace."

Attempting to address a perception that politics were taking precedence over science, planners of the 1975 Nobel Conference sought to return it to its roots. "The Future of Science," however, also raised another landmark in Nobel Conference history.

Celebrating the first ten years of the conference would necessarily involve a measure of pomp and circumstance. How big that measure would be came as a happy surprise to all 3200 in attendance: twenty-seven Nobel laureates came to Gustavus for the conference. The next several conferences focused on physics, chemistry and biology, with Nobel laureates Murray Gell-Mann, Max Delbruck and Tjalling Koopmans presenting talks. In 1979, Nobel Conference XV grasped an invisible hand and stepped into a new area: economics. "The Future of the Market Economy" aroused controversy through calls from the United States (Kenneth McLennan), Canada (Richard Lipsey) and Sweden (Baron Stig Ramel, President of the Nobel Foundation) for a more free-market economy.

That controversy was minor, however, when compared to the one joined at Nobel Conference XVIII in 1982. With "Darwin's Legacy" as the topic, panelists including Richard Leakey, Stephen Jay Gould, Peter Medawar and Edward O. Wilson took on both the creationist attack on the theory of evolution and traditional understandings of Darwin's work.

The resulting brouhaha, with angry letters again fired off to newspapers and journals, did not dissuade anyone from attending the following year's conference, "Manipulating Life." Some 4500 people, including 20 visiting scholars from China, heard Lewis Thomas, Nobel Laureate Christian Anfinsen and four other authorities discuss the latest advances in genetic research.

The most recent Nobel Conferences have seen representatives of as many as 95 colleges and universities and over 115 high schools in attendance. Economist James Buchanan received the 1986 Nobel Prize only a week after his appearance at Nobel Conference XXII, "The Legacy of Keynes." The 1988 conference, "The Restless Earth," was the first devoted to the science of geology.

This year's Conference, XXV, "The End of Science," marked the twenty-fifth anniversary of the Conference that had begun with "Genetics and the future of Man." This unique theme was the construction of three Gustavus professors, John Olson, Bill Dean and Dick Elvee. They sketched it out while eating pork ribs in a hot sauce at Hardy's in Mankato, Minnesota. They thought it rather tongue-in-cheek but also relevant. Here the college was celebrating 25 years of the Nobel Conference, yet it questions the staying power of science as we know it.

Their provocative letter of invitation proved to be both the draw and the catalyst for one of the most interesting conferences in the Nobel history. The letter as follows was sent to Physicist, Sheldon Glashow; Biologist, Gunther Stent; Philosophers, Sandra Harding, Mary Hesse and Ian Hacking, and Historian Gerald Holton. The invitation read:

The committee's reasoning as regards the theme is as follows: "Science as a unified, universal, objective endeavor is currently being questioned. In recent times science has come to be understood as a more subjective and relativistic societal factors. Thus science is currently re-examining itself as the product of such things as paradigmatic focuses, ideological struggles and the basic instrument of power.

For example, whereas previously science was thought to be "tainted" if it was under the direction of culture/government/society, such as the blatant Stalinist intervention in Russian science, as a whole, per Feyerabend, is understood as selective and "repressive!" It is with this new understanding of science that science as science is seen as coming to an end; i.e., science is no longer a fortress of objectivity.

We are also grappling with grave epistemological issues. If science is not a reflection of some extra-historical, external, universal laws, but is social, temporal and local, and if it evolves just as surely as a species evolves, then there is no way of speaking of something real behind science that science merely reflects. Science is what we know about nature, and what we know is a function of time, place and the accidents of communities and their conversation.

But this leaves unanswered the question, What does science account for, if not eternal and universal laws? More than anything else, science is a sequence of interpretations of nature; however, the interpretations are creations of interpreters of nature, and to a large extent science is an account of the imagination of the interpreters (observers) of nature. This means not only that the history of the intellectual endeavor that for 400 years has been called science is, to a large extent, a creature of the scientist's subjectivity but also that those objective regularities that the scientists interpret themselves are composed of observer responses to the world. Those responses give the natural world whatever new ingredients it gets, whether the response be that of an atom, an animal, or an animated scientist. In other words, the scientific object is created by subjects just as much as the scientific subject.

With the change that finishes off science as an objective, logocentric discipline, comes the question, What, given this development, should science mean to society? It can no longer mean, as it meant to modernism, the stable guide in a world of humanistic instability. Is this new understanding of science forecasting THE END OF SCIENCE?"

The Conference took place on October 3 and 4, 1989, at Gustavus Adolphus College, in the presence of 4000 attendees. The six world-class scientists, philosophers, intellectuals invited to speak were brilliant in script and delivery, in dialogue and repartee and in final commentary. Their addresses, remarks and commentaries are contained in this book.

Sandra Harding argued that modern science, like religion in the Victorian Age, is suffering grave doubts. The idea of a unified science--redemptive, uncovering objective reality--is a naive fantasy. Rather, science is no longer a monolithic and male-dominated, bourgeois pursuit, but has become pluralistic and changeable, its conclusions influenced by ideologies ranging from Marxism to feminism to pacifism.

In brief, Sheldon Glashow, Nobel laureate (1979) argued that the power of science has not been diminished or fractured. Despite the attempts of ideological groups to usurp it, science remains the one bold hope for the resolution of earthly problems such as pollution and illiteracy. Moreover, science is a tool for understanding the hidden secrets of the universe.

Ian Hacking argued that science is no longer a model of all civilized knowledge, nor should it be. It suffers from false portrayals. Humanists, abetted by pure theorists, cling to the picture of science as an intellectual adventure striving to find the one ultimate theory of everything. Yet we now profess other values of accommodation, variety, choice. The sciences are not one single thing at all. They have become as multifloriate as the humanities. They have stability without unique foundations, sharing without complete commensurability, competition without banal subjectivity, and realism without a fantasy of ultimate reality. Their models of aspects of the world may be constructions, but they are subject to rigorous constraints. They are less engaged in representing reality than in intervening in it to create new phenomena. The loose unities among the sciences are genuine, but are less a result of theory than of a motley of practices, ranging from fast computation to off the shelf instruments that are passed from one discipline to another.

Mary Hesse argued that the pretensions of science to be the unique road to knowledge have come to a finis. The proper telos of natural science is limited: it is the discovery of local natural regularities, and their exploitation under social and human constraints that science itself cannot deliver. Scientific theories are not literally true descriptions of states of affairs, any more than are perceptual models, or imaginative metaphors, narratives or myths. Scientific knowledge is local and approximate; its linguistic representations require a theory of meaning and truth which is not propositional; its logic is not deductive. Hesse thinks that Cognitive Science and Artificial Intelligence can tell us a great deal about how to reconcile mental construction and objectivity. Objective knowledge is primarily a matter of "know-how": it is the skill of combining experience, perception, interpretation and action to realize human purposes. "Know-that" involves language, and language necessarily loses information compared with know-how, because language is not only a finite system, but a very small system compared with experience.

Gerald Holton argues that current worries about the inadequacy of science to explain our expanding universe are nothing new. But it is still our best tool, and without it, we would face almost unimaginable catastrophes. The current preoccupation with the decay of science is neither a contradiction nor a novelty. It was raised at the end of the 18th century when our limited knowledge of physics could not provide answers to newly-found phenomena. The bankruptcy of science was proclaimed until that same flawed science discovered quantum theory and relativity, launching us into new frontiers of knowledge. Two schools of thought arise: the linearists who believe science advances in a linear manner, and the cyclicists who argue that scientific activities arise and fall in a circular, evolutionary form. Spengler is the cyclist who argues that near the end of our modern civilization, there evolves a form of lonely romanticism, then an intellectualism which turns its attentions to abstractions. This is the winter season when the primacy of the soul is pre-empted by intellectualism. As science is at its zenith it plants the seeds of its own undoing, and those self destructive tendencies eventually kill it. But science today shows no signs of perishing. Spengler's views are not flawless. Holton cites Einstein as a linearist who thought the ultimate unification of science in the goal of knowing natural processes would be unending. Einstein believed the whole spectrum of theories that appear as mutually exclusive are actually complementary necessities that make up the theory of pre-established harmony. We must harness the impulses to transcend our present understandings, and find certifiable truths.

Gunther Stent argued that three cognitive limits of science have come into view in the present century. A semantic limit surfaced with the dramatic developments of modern physics, when it turned out that our intuitive categories, such as object, time and space are inadequate; to fathom nature in its deepest aspects. A structural limit emerged in both physical sciences (such as in meteorology and astrophysics) and social sciences (such as economics and sociology) upon the attempt to understand phenomena whose critical data are hidden in a fog of noise. Finally, a subjective limit appeared with the study of highly complex natural phenomena, to the interpretation of which scientists must bring their personal preunderstanding.

Why Physics is a Bad Model for Physics

SANDRA HARDING

The sciences incorporate both liberatory and oppressive tendencies and have done so since their origins. The new sciences of the 17th century decentered our species from its unique location in a universe described by Christian thought. They said instead that humans are located on an otherwise ordinary planet circling around an unremarkable sun in an insignificant galaxy, and, further, that the earth and the heavens are made up of the same kinds of materials and moved by the same kinds of forces. Thus those new sciences gave anti-aristocratic messages; they implied that nature does not determine any higher or lower stations in life, nor does nature make higher or lower human "natures." In this way they undermined belief in the natural legitimacy of royalty and aristocracy. Moreover, they were epistemologically anti-authoritarian and participatory. "Anyone can see through my telescope," said Galileo, and then reason to the conclusions of the new sciences. We are used to thinking of this particular set of social values that are carried by modern science in contradictory ways. On the one hand, they are not social values at all, since even though science incorporates them it can still attain value-neutrality. On the other hand, these values are thought to be so constitutive of science that someone who criticizes science is thought to be against reason, progress and democracy.

However, the new sciences carried other kinds of social values also, for they provided resources for a new social class to assert its legitimacy over others. This class had interests in owning land and developing its resources (its ores, plants, animals, and the peasants who also belonged to the land) for its own benefit, in using warfare to obtain access to land and its resources, and in legitimating only its own activities and achievements as what everyone should recognize as civilization. These interests found a ready companion in the focus of the new sciences on the materiality of the world, on developing more efficient ways to dominate nature, on the value of technological "progress," and on the legitimacy and usefulness of universal laws. Modern Western science was constructed within and by political agendas that contained both liberatory and oppressive possibilities.

Present-day science, too, contains these conflicting impulses. The anti-democratic impulses are not only morally and politically problematic; they also deteriorate the ability of the sciences to provide objective, empirically defensible descriptions and explanations of the regularities and underlying causal tendencies in nature and social relations. One way to focus on this problem is to discover that we do not have a conception of objectivity that enables us to distinguish the scientifically "best descriptions and explanations" from those that fit most closely (intentionally or not) with the assumptions that elites in the West do not want critically examined. Without such a strong criterion of objectivity, science can easily become complicitous with the principle that "might makes right," whether or not anyone intends this complicity. The ethics and rationality of science are intimately connected.

Here I shall focus on some popular but false beliefs about the sciences that block the development of democratic and objective impulses in science. I frame the following discussion in terms of the project of separating science from its complicity with sexist and androcentric agendas--a project that I have had a hand in developing, though I can not here take the space also to discuss these feminist critiques.[1] However, the false beliefs to be discussed have also supported complicity of the sciences with political agendas that are racist, Eurocentric and bourgeois.[2]

Science Without the Elephants. Are feminist criticisms of Western thought relevant to the natural sciences? Many people who understand the importance of feminist criticisms for removing sexist bias from the social sciences and humanities have difficulty appreciating their relevance to the natural sciences.[3] "Of course, there should be more women in science, mathematics and engineering--and the good ones will rise to the top," they say. "Moreover, it is not at all good that some technologies and applications of natural science have been dangerous to women; policy-makers should take steps to eliminate these misuses and abuses of the sciences. But the logic of research design and the logic of explanation in the physical sciences are fundamentally untouched by the feminist criticisms and will necessarily remain so. This is because the logic of research and of explanation, and the cognitive, intellectual content of natural science's claims--'pure science'--cannot be influenced by gender."

This argument will not stand up to scrutiny. It is grounded not only in underestimations of the pervasiveness of gender relations--relations that appear not only between individuals,but also as properties of institutional structures and

---

[1]See, e.g., Discovering Reality: Feminist Perspectives on Epistemology, Metaphysics, Methodology and Philosophy of Science, ed. S. Harding and M. Hintikka (Dordrecht: Reidel Publishing Co., 1983); The Science Question in Feminism (Ithaca: Cornell University , 1987); Sex and Scientific Inquiry, ed. S. Harding and J. O'Barr (Chicago: University of Chicago Press, 1987), Feminism and Methodology: Social Science Issues, ed. S. Harding (Bloomington: Indiana University Press, 1987).

[2]Important questions about the consequences for the Third World of the development and character of Western sciences are raised by Walter Rodney, How Europe Underdeveloped Africa (Washington, D.C.: Howard University Press, 1982); Susantha Goonatilake, Aborted Discovery: Science and Creativity in the Third World (London: Zed Books, 1984); Ivan Van Sertima, Blacks in Science: Ancient and Modern (New Brunswick: Transaction Books, 1986); V. Y. Mudimbe, The Invention of Africa: Gnosis, Philosophy, and the Order of Knowledge (Bloomington: Indiana University Press, 1988).

[3]The feminist literature that is relevant to the natural sciences has rapidly expanded. The most recent comprehensive bibliography (it includes social science critiques) is "Feminist Critiques of Science", A. Wylie and K. Okruhlik, Resources for Feminist Research (forthcoming 1989). A good place to start is the Harding and L'Barr collection, Op. Cit. that contains reprints from the Journal, Signs. It begins with a review essay by Londa Schiebinger, "The History and Philosophy of Women in Science: A Review Essay" (p. 7-34), and also includes a number of the most widely-cited papers in the field.

of symbolic systems[4]--but also in false beliefs about the natural sciences. Because of these false beliefs, it is difficult to make sense of many aspects of science and society. One can think of these false beliefs as extraneous elements in meta-theories of science. If we remove them, we can begin to understand aspects of science that appear inconsistent or inexplicable as long as we hold these beliefs.

"Physics" in the title of this essay refers to a certain image of science that is full of these mystifying beliefs. This "physics" is magical; it is like the ancient image of a column of elephants holding up the earth. The logic of the column of elephants--"You can't fool me, young man: it's elephants all the way down," as the punch line to the old joke goes--prevents the observer from asking questions that would quickly come to mind were the elephants not so solidly in view. Physics is to "physics" as a satellite photo of the earth is to a picture of the earth balanced on top of a column of elephants. We can understand physics without "physics."[5]

The reader should be assured that I do not intend to throw out the baby of science along with its bathwater of false views about science.[6] Instead, the concern here is to separate false beliefs about the sciences from ones that are conducive to empirically, theoretically, and politically more adequate sciences. The intent is to identify more carefully where the baby ends and the bathwater begins. There are some <u>causes</u> of scientific beliefs and practices that are to be found outside the consciousnesses of individual scientists; that is, they are not <u>reasons</u> for the acceptance or rejection of these beliefs and practices. Our society is permeated by forms of scientific rationality. It is in just such a society that there is a deep resistance to understanding the institutional practices of science and how these shape the activities and consciousnesses of scientists as well as of the rest of us. From the perspective of the democratic tendencies within science, that resistance is irrational. This irrational resistance frames discussions so that it is difficult for people to understand their own activities and why some of the choices with which they are confronted are such limited and narrow ones. The false beliefs to be examined below serve to hide that irrationality from critical scrutiny.

---

[4]See p. 52-56 of <u>The Science Question in Feminism Op. Cit.</u> for discussion of these 3 sometimes conflicting manifestations of gender relations.

[5]The skeptic with whose response to feminist criticism I began is already far more open to feminism than his even more conventional colleagues. The "super-conventionalists" believe, though he does not, that if feminism is relevant to the social sciences, it is only because the latter do not really deserve to be regarded as reliable modes of gaining knowledge; they certainly should not be thought of as sciences. In the social sciences, values frequently shape the results of research because of assumptions and procedures that are contrary to the most rigorous scientific practices. Some super-conventionalists believe that insofar as feminist criticism points out these social values, it can make the positive contribution of showing the need for more rigorous adherence to the existing norms of scientific inquiry. Many others claim that feminism does not in fact have such a positive effect on social science at all; they think it simply substitutes one set of social biases for another.

[6]Contrary to the recommendations of, e.g., Sal Restivo, "Modern Science as a Social Problem," <u>Social Problems</u> Vol 35, no. 3, p. 206-225 (1988).

Some readers will think I am criticizing a straw figure. They will find it convenient to think that only positivist tendencies that are no longer fashionable are the reasonable target of these criticisms. I can not here detour to define positivism and debate its influence. But it is simply a fact widely-recognized in the social studies of science that while fewer scientists, philosophers and social scientists who model their work on the natural sciences are as openly enthusiastic about positivism compared to the situation 40 and more years ago, most of these people still happily embrace fundamental assumptions of positivism.

1. "Feminism is about people and society; the natural sciences are about neither. So feminism can have no relevance to the logic or content of the natural sciences." One line of thinking behind this argument is that researchers are far more likely to import their social values into studies of other humans than into the study of stars, rocks, rats or trees. And it is absurd, such a conventionalist will argue, to imagine that social values remain undetected in the studies of the abstract laws that govern the movements of the physical universe. Scientific method has been constructed exactly to permit the identification and elimination of social values in the natural sciences. Practicing scientists and engineers often think the discussions of objectivity and method by philosophers and other non-scientists are simply beside the point. If bridges stand and the television set works, then the sciences that produced them must be objective and value-free-- that's all there is to the matter.

One could begin to respond by pointing out that evolutionary theory, a theory that is about all biological species and not just about humans, clearly "discovered" secular values in nature, as the Creationists have argued. It also "discovered" bourgeois, Western and androcentric values, as many critics have pointed out.[7] Moreover, the astronomy and physics of Newton and Galileo, no less than those of Aristotle and Ptolemy, were permeated with social values, as indicated earlier. Many writers have identified the distinctively Western and bourgeois character of the modern scientific world view.[8] Some critics have detected social values in contemporary studies of slime mold, and even in such abstract sciences as relativity theory and formal semantics.[9] Conversationalists

---

[7]E.g., Stephen Jay Gould, The Mismeasure of Man (New York: W. W. Norton, 1981); Ruth Hubbard, "Have Only Men Evolved?" in Harding and Hintikka, Op. Cit.

[8]See Leszek Kolakowski, The Alienation of Reason: A History of Positivist Thought, (New York: Doubleday, 1968); Carolyn Merchant, The Death of Nature: Women, Ecology and the Scientific Revolution (New York: Harper & Row, 1980); Alfred Sohn-Rethel, Intellectual and Manual Labor, (London: Macmillan, 1978); Margaret C. Jacob, The Cultural Meaning of The Scientific Revolution (New York: Alfred A. Knopf, 1988); W. Van den Daele, "The Social Construction of Science," in E. Mendelsohn, P. Weingart, R. Whitley, eds., The Social Production of Scientific Knowledge (Dordrecht: Reidel, 1977); Morris Berman, The Reenchantment of the World (Ithaca: Cornell University Press, 1981).

[9]Respectively, Evelyn Fox Keller, "The Force of the Pacemaker Concept in Theories of Aggregation in Cellular Slime Mold," and "Cognitive Repression in Contemporary Physics," both in Reflections on Gender and Science (New Haven: Yale University Press, 1985); Paul Forman, "Weimar Culture, Causality, and Quantum Theory, 1918-1927: Adaptation by German Physicists and Mathematicians to a Hostile Intellectual Environment," Historical Studies in the Physical

respond to this by digging in their heels. They insist on a sharp divide between pre-modern and modern sciences by claiming that while medieval astronomy and physics were deeply permeated with the political and social values of the day, the new astronomy and physics were (and are) not; this is exactly what distinguishes modern science from its forerunners. As one historian of science said back when he was such a conventionalist, the world view characteristic of medieval Europe was much like that of "primitive societies" and children:

> the world view of primitive societies and of children tends to be animistic. That is, children and many primitive peoples do not draw the same hard and fast distinction that we do between organic and inorganic nature, between living and lifeless things. The organic realm has a conceptual priority, and the behavior of clouds, fire, and stones tends to be explained in terms of the internal drives and desires that move men and, presumably, animals."[10]

The conventionalist fails to grasp that modern science has been constructed by and within power relations in society, not apart from them.[11] The issue here is not how one scientist or another used or abused social power in doing his science. Rather we should ask where the sciences, their agendas, concepts and consequences have been located within particular currents of politics. How have their ideas and practices advanced some groups at the expense of others? Can sciences that avoid such issues understand the causes of their present practices, of the changing character of the tendencies they seem to "discover in nature" in different historical settings?

Even though there are no complete, whole humans visible as overt objects of study in astronomy, physics and chemistry, one cannot assume that no social values, no human hopes and aspirations, are present in human thought about nature. Consequently, feminism can have important points to make about how gender relations have shaped the origins, problematics, decisions about what to count as evidence, social meanings of nature and inquiry and consequences of scientific activity. In short, we could begin to understand better how social projects can shape the results of research in the natural sciences if we gave up the false belief that because of their non-human subject matter, the natural sciences can produce impartial, disinterested, value-neutral accounts of a nature that is completely separate from human history.

    2. "Feminist critics claim that a social movement can be responsible for generating empirically more adequate beliefs about the natural world. But only

---

Sciences, Vol. 3 (1971); Merrill B. Hintikka and Jaakko Hintikka, "How Can Language Be Sexist?" in Harding and Hintikka, eds., Op. Cit.

[10]Thomas Kuhn, The Copernican Revolution (Cambridge: Harvard University Press, 1957), p. 96.

[11]See Joseph Rouse, Knowledge as Power: Toward a Political Philosophy of Science (Ithaca: Cornell University Press, 1987); Merchant, Op. Cit.; W. Van den Daele, Op. Cit.; my discussion in Chapters 8 and 9 of The Science Question in Feminism, Loc. Cit.

false beliefs have social causes. Whatever relevance such critics have to pointing out the social causes of false beliefs, feminism cannot generate 'true beliefs.'" This claim assumes that no social science findings could be relevant to our explanations of how the best, the empirically most supported (or least refuted) hypotheses arise and gain scientific legitimacy. Some conventionalists will agree that the social sciences can tell us about the intrusion of social interests and values into research processes that have produced false beliefs. We want to know why phlogiston theory, phrenology, Nazi science, Lysenkoism and creationism were able to gain a legitimacy and respect that they should not have had; the causes of their illegitimate status are to be found in social life. There is a worthy task for sociologists and historians. But the content of "good science" has no social causes but only natural ones, according to the conventionalist. It is a result of the way the world is, the way our powers of observation and reason are, and of bringing our powers of observation and reason to bear on the way the world is. Consequently, the most widely accepted natural science claims require no causal accounts beyond the reasons scientists could give for their own cognitive choices.

Supporting this view of the asymmetry of causal explanations of belief is a long tradition in epistemology that has been criticized in recent decades by sociologists of knowledge.[12] They argue that it is simply a prejudice of philosophers to hold that the beliefs a culture regards as legitimate should uniquely be excepted from causal social explanations. To hold such a position is to engage in mysticism: it is to hold that the production of scientific belief, alone of all distinctively human social activities, has no social causes. Instead, they argue, a fully scientific account of belief will seek causal symmetry; it will try to identify the social causes (as well as the natural ones) of the best as well as the worst beliefs.

This sociological account is flawed in a variety of ways. For one thing, these writers appear to exempt their own claims from the causal accounts they call for elsewhere, in this and other ways adopting still excessively positivist conceptions of scientific inquiry.[13] Moreover, this account appears to reduce scientific claims simply to beliefs that happen to be socially acceptable. It offers no way to talk about the natural constraints within which historically distinctive scientific accounts are produced.[14] But we do not have to replicate the limitations of these sociological accounts; we are not forced to the functionalism and

---

[12]David Bloor, Knowledge and Social Imagery (London: Routledge & Kegan Paul, 1977); Barry Barnes, Interests and the Growth of Knowledge (Boston: Routledge and Kegan Paul, 1977); Karin Knorr-Cetina, The Manufacture of Knowledge (Oxford: Pergamon, 1981); Karin Knorr-Cetina and Michael Mulkay, eds., Science Observed: Perspectives on the Social Study of Science (Beverly Hills: Sage, 1983.)

[13]See, e.g., Bloor, Op. Cit, p. 142ff. An attempt to remedy this situation by pursuing to its amusing though disastrous end the embrace of relativism required by the logic of the "strong programme" in the sociology of knowledge can be seen in Steve Woolgar, Knowledge and Reflexivity (Beverly Hills: Sage, 1988).

[14]Hilary Rose, "Hyper-reflexivity: A New Danger for the Counter Movements," in Helga Nowotny and Hilary Rose, Counter-Movements in the Sciences: The Sociology of the Alternatives to Big Science (Dordrecht: D. Reidel Publishing Co., 1979.)

relativism that plagues these otherwise illuminating analyses. We can hold that our own (true! Or, at least, less false) account also has social causes--that, for example, changes in social relations have made possible the emergence of modern science's distinctive intellectual and political trajectory as well as of feminisms's. These histories leave their fingerprints on the cognitive content of science no less than of feminism.[15] Moreover, we can insist that the identification of social causes for the acceptance of a belief does not exclude the possibility that such a belief also does match the world in better ways than its competitors. That is, we can hold that certain social conditions make it possible for humans to produce more reliable explanations of patterns in nature just as other social conditions make it very difficult to do so.

If the objection above to feminist accounts of the social causes of "true belief" were reasonable, one would have to criticize on identical grounds the new histories, sociologies, psychologies, anthropologies and political economies of science. A wide array of studies have shown the politics within which modern scientific knowledge has been constructed. Eliminating the idea that only false beliefs can have social causes--this "elephant"--makes it possible to provide more coherent accounts of what actually has contributed to the growth of knowledge in the history of the sciences. It makes it possible to understand feminism as able to advance knowledge not only by debunking false beliefs, but also by helping to change social conditions into those more conducive to the recognition of less partial and distorting belief.

3. "Science is fundamentally only the formal and quantitative statements that express the results of research; (and/or) science is a unique method. If feminists do not have alternatives to logic and mathematics or to science's unique method, then their criticisms may be relevant to sociological issues, but not to science itself." Galileo argued that "nature speaks in the language of mathematics," so that if we want to understand her, we must learn to speak her language. Some conventionalists have understood this to mean that "real science" is only the formal statements that express such laws of nature as those discovered by Newton, Boyle and Einstein. There are a number of problems with such claims.[16] While there can appear to be no social values in results of research that are expressed in formal symbols, formalization does not guarantee the absence of social values. For one thing, historians have argued that the history of mathematics and logic is not merely an external history about who discovered what when. They claim that the very general social interests and preoccupations of a culture can appear in the forms of quantification and logic that its mathematics uses. A number of the most distinguished mathematicians have concluded that the ultimate test of the adequacy of mathematics is a pragmatic one: does it work to do what it was intended to do? For example, the need for

---

[15]See, e.g., my "Why Has the Sex/Gender System Become Visible Only Now?," in Harding and Hintikka, eds. Op. Cit, Van den Daele, Op. Cit, and Edgar Zilsel, "The Sociological Roots of Science," American Journal of Sociology, Vol. 47 (1942).

[16]This section repeats the arguments made in Chapter 2 of The Science Question in Feminism Op. Cit.

accurate measurement requires the invention of mathematical notions that conflict with ones useful for counting.[17]

Moreover, formal statements require interpretation in order to be meaningful. The results of scientific inquiry can count as results only if scientists can understand what they refer to and mean. Without decisions about their references and meanings, they cannot be used in any useful way, such as for prediction or to stimulate future research. As is the case with social laws, the referents and meanings of the laws of science are continually extended and contracted through decisions about the circumstances in which they should be considered to apply. What is to count as a case of "1", or of "2", of "+" or of "-"? Making such decisions is a process of social interpretation.

There is also the fact that metaphors have played an important role in modeling nature and specifying the appropriate domain of a theory.[18] To take a classic example, "nature is a machine" was not just a useful heuristic for explaining the new Newtonian physics. It was an inseparable part of that theory, one that created the metaphysics of the theory and showed scientists how to extend and develop it. Thus social metaphors provided part of the evidence for the claims of the new sciences; some of their more formal properties still appear as the kinds of relations modelled by the mathematical expressions of the natural sciences. They were not only "outside" the process of testing hypotheses; they were also "inside" it. The social relations of the period that both made possible and were in turn supported by the machines on which Newton's mechanistic laws were modelled functioned as--were--part of the evidence for Newtonian physics. Giving up the belief that science is really or fundamentally only mathematical statements is necessary if we are to begin to explain the history and practices of science. Insistence on this belief is a way of irrationally restricting thought.

If science is not reducible to its formal statements, is it reducible to its method? This is an equally problematic claim. Contemporary physicists, ethologists and geologists collect evidence for or against hypotheses in ways different from how medieval priests collected evidence for or against theological claims. However, it is difficult to identify or state in any formal way just what it is that is unique about the scientific methods. For one thing, different sciences develop different ways of producing evidence, and there is no clear way to specify what is common to the methods of high energy physics, ethology, and plate tectonics. "Observing nature" is certainly far too general to specify uniquely scientific modes of collecting evidence; gatherers and hunters, pre-modern farmers, ancient seafarers, and mothers all must "observe nature" carefully and continuously in order to do their work. These examples also show that linking prediction and control to the observation of nature are certainly not unique to science, since they are also crucial to gathering and hunting, farming, navigation, and child care. Scientific practices are common to every culture. Moreover, many phenomena of interest to science cannot be controlled though they can be predicted and explained--for example, the orbit of the sun and the location of

---

[17]Morris Kline, <u>Mathematics: The Loss of Certainty</u> (New York: Oxford, 1980); Bloor <u>Op. Cit.</u>

[18]Mary Hesse, <u>Models and Analogies in Science</u> (Notre Dame: University of Notre Dame Press, 1966); Merchant <u>Op. Cit.</u>; see also my discussion of Hesse's conclusions in Harding 1986, <u>Op. Cit.</u>, p. 233-9.

fossils. And prediction alone is possible on the basis of correlations that in themselves have little or no explanatory value.

Philosophers and other observers of science have argued for centuries over whether deduction or induction should be regarded as primarily responsible for the great moments in the history of science,[19] but it is obvious that neither is unique to modern science: infants and dogs regularly use both. It may be futile to try to identify the distinctive features of knowledge-seeking that will exclude from the ranks of people who should be counted as scientists mothers, cooks, or farmers, but include highly trained but junior members of, say, biochemical research teams. This is even more true in a society such as ours where scientific rationality has permeated child care, cooking and farming.

One might try to defend the idea that the important feature of scientific method is science's critical attitude.[20] That is, scientific method is fundamentally a psychological stance. In all other kinds of knowledge-seeking, this line of argument goes, we can identify assumptions that are regarded as sacred or immune from refutation. Only modern science holds all of its beliefs open to refutation. However, this proposal is not supported by the history, present practices, or leading contemporary meta-theories of science. On the one hand, assumptions that are held immune from criticism--either on principle or inadvertently--are never absent from the sciences. The history of science shows again and again that scientists and science communities make unjustified assumptions, and that they are loath to critically examine the hypotheses in whose plausibility they have invested considerable time, energy and reputation. Moreover, there are some beliefs that we could call constitutive of science in the sense that they can be questioned only at the risk of creating skepticism about the whole enterprise of science. Examples would be the idea that all physical events and processes have causes, even if we can't always know what they are, and that it is a good thing to know more about nature. Furthermore, everyone understands that there must be many scientific assumptions that are questionable in principle, but that cannot all be questioned simultaneously if research is to occur at all. Moreover, Kuhn proposed that a field of inquiry only really becomes a science when it decides to accept some set of beliefs as "not to be contested," and makes these the assumptions that define the field.[21] Others point to the necessarily unquestioned "background assumptions" or "auxiliary hypotheses" that inevitably hover behind every hypothesis that is being tested. These include optical theories, beliefs about how the testing and recording instruments work, assumptions about which are the significant variables, about what can count as a repeated observation or experiment, etc.

---

[19]See S. Harding, ed., Can Theories Be Refuted? Essays on the Duhem-Quine Thesis (Dordrecht: Reidel, 1976).

[20]Karl Popper, Conjectures and Refutations: The Growth of Scientific Knowledge, 4th ed. (London: Routledge & Kegan Paul, 1972); Robin Horton, "African Traditional Thought and Western Science," Pts. 1 & 2, Africa 37 (1967).

[21]Thomas Kuhn, The Structure of Scientific Revolutions, 2nd ed. (Chicago: University of Chicago Press), 1970. (This line of thought leads Kuhn to dubious claims about how to create true sciences, as we shall see.)

Furthermore, Western science is not the only domain of critical thought. Everyone must have a critical attitude toward a good number of beliefs if they are to survive the vicissitudes of nature and social life. It is part of the ethnocentrism of the West to assume that only practitioners of Western scientific rationality exercise critical reason. Feminists and the working class have also questioned the assumption that critical reason is the talent only (or even!) of the dominant groups.[22]

The ideas that science is really or fundamentally formal statements or a distinctive method are extraneous beliefs that block our ability to describe and explain the workings of modern Western science. Science has many interlocking practices, products, referents and meanings. It is a cumulative tradition of knowledge. It is an "origins story" that is a fundamental part of the way certain groups in the modern West identify themselves and distinguish themselves from others. It is a metaphysics, an epistemology, an ethics and a politics that is compatible with the agendas of modern liberal states, capitalism, and Protestantism. It is not compatible with every kind of government, economy, or moral and religious tradition--at least not in the sciences' modern Western forms. Some have pointed out that it not only has become a religion for many, but that it intends to hold the place of a religion. What else, they ask, could one conclude about its insistence on its own absolute authority, on its "monologue" form, on its inherent moral good, its intolerance of criticisms from "outside," and its intended use to define the borders of "civilization." It attempts to hide its religious character by distancing itself from religion. It is a social institution with complex rituals and practices that both reflect and shape social relations in the cultures in which it exists. It is both the producer and the beneficiary of technological invention. It is a major factor in the maintenance and control of production and, increasingly, reproduction.

It is striking to contrast this array of descriptions of "what science is" with the restricted range upon which conventionalists insist. These false beliefs block our ability to explain how science works.

4. "Technologies and applications of science are not part of science proper. So feminist criticisms of the misuses and abuses of the sciences (such as of the proliferation of dangerous reproductive technologies) challenge only public policy about science, not science itself." Preceding discussions indicate why this is a distorted representation of science, technology and the relations between them. Whatever was true in the past, it is difficult now to identify anything at all that it is reasonable to count as pure science. Is this too strong a claim? Let us see. Science makes use of technological ideas and artifacts at least as much as the reverse. Moreover, just because scientific ideas may not result in any immediate application doesn't show that those ideas or the practices producing them are "pure" since they may very well still be permeated with values. After rethinking

---

[22]See, e.g., J. E. Wiredu, "How Not to Compare African Thought with Western Thought," in African Philosophy: An Introduction, 2nd ed., ed. Richard A. Wright. (Washington, D.C.: University Press of America, 1979); Genevieve Lloyd, The Man of Reason: "Male" and "Female" in Western Philosophy (Minneapolis: University of Minnesota Press, 1984); Carol Gilligan, In a Different Voice: Psychological Theory and Women's Development (Cambridge: Harvard University Press, 1982); M. G. Belenky, B. M. Clinchy, N. R. Goldberger, J. M. Tarule, Women's Ways of Knowing: The Development of Self, Voice, and Mind (New York: Basic Books, 1986).

the complex relationship between sciences and technologies, many observers have concluded that science is politics by other means. It is more than that, but it is that, too.

Everyone is willing to acknowledge that scientific research makes possible new technologies and applications of science. Science produces information that can be applied in the social world and used to design new technologies. This is not thought to threaten the purported purity of science since it is not scientists but policy makers who actually decide to construct the technologies and carry out the new applications of scientific information. "You can't infer an 'ought' from an 'is,'" as philosophers like to say. Deciding what we ought to do with the information science provides is supposed to be a separate process from producing the information in the first place. According to this way of thinking, it is policy makers who should be held responsible for the misuses and abuses of the sciences and their technologies--not scientists or the sciences themselves.

Because two distinct groups of people have responsibility for the two kinds of decisions, it is easier to think that the technologies and their sciences must be conceptually and politically separate. Scientists in universities produce the information, and scientists in industry, the military and the government make the decisions about what information is to be disseminated and how it is to be used.[23] But this division of labor does not have the consequences its defenders suppose. It simply makes it difficult for scientists in universities to be able to explain their own activities in a plausible way--that is, to give the kind of causal account of science that scientists recommend we give about everything else. Their explanations of their activities do not maximize coherence, generality, simplicity, fit with empirical evidence, etc.

In the first place, some "is's" in practice insure "oughts." For example, in a racist society, "pure descriptions" of racial difference have little chance of functioning as pure information. One can be confident that racist assumptions will markedly narrow the range of "reasonable" applications of such "information." Moreover, the very concern with racial difference in such a culture cannot be free of race value.[24] The scientific reports can be as value-neutral as possible in the sense that they describe only difference, not inferiority and superiority, and make no recommendations for social policy. But it is exactly this kind of research that one can reasonably predict will be used for racist ends (intentionally and not) in a race-stratified society. (This is not an argument against doing such research, but against the refusal to state and discuss publicly the political interests in and possible consequences of the research.)

Does it continue to make sense to refer to this kind of research as objective inquiry when everyone has a stake in its outcome? Moreover, as social scientists have pointed out, it cannot be value-free to describe such social events as poverty, misery, torture or cruelty in a value-free way. When faced with such phenomena, every statement counts as either for or against; there is no possibility of a third stance that is value-free. The use of objective language to describe such events

---

[23]See, e.g., Paul Norman, "Behind Quantum Electronics: national Security as Bases for Physical Research in the U.S. 1940-1960," in Historical Studies in Physical and Biological Sciences, v. 18 (149-229), 1987.

[24]See Gould Op. Cit.

results in a kind of pornography; the reader, the observer, consumes for his/her own intellectual satisfaction someone else's pain and misfortune. On a related topic, some critics argue that scientific method does not appear to provide any criterion for distinguishing whether certain procedures on humans should subsequently be referred to as scientific experiments or torture.[25]

Defenders of pure science frequently appear to be arguing that ignorance of the consequences of one's scientific behaviors should be counted as evidence for one's objectivity. But since the law finds avoidable ignorance culpable, why shouldn't science? Of course, no one can guarantee the good consequences of all of one's decisions or perhaps even of any of them. But the point I am making is a different one: why shouldn't it be regarded as culpable to refuse to consider the consequences of one's acts, as this insistence on the possibility of a separation between pure and applied science directs us to do? The "innocence" of science communities--our "innocence"--is extremely dangerous to us all. Perhaps people who have exhibited tendencies toward such innocence should not be permitted to practice science or construct meta-theories of science; they are a danger to the already disadvantaged--perhaps even to the species! Why shouldn't we regard ignorance of the reasonably predictable consequences of one's scientific behaviors not as evidence of the objectivity of that research, but of incompetence to conduct it? I am putting this issue in terms of moral responsibility although it is fundamentally a political issue: how is modern Western science constructed by class, race and gender struggles? But claiming individual moral responsibility can be a powerful motive for political change.

It is less widely recognized that the technologies science uses in its research processes themselves have political consequences. For example, the use of the telescope moved authority about the heavens from the medieval church to anyone who could look through a telescope. The introduction of complex diagnostic technologies in medical research moves authority about the condition of our bodies from us to medical specialists. In practice, it tends to move this authority even away from physicians' and toward lab technicians. These are not trivial involvements of science in political interests and values. Not all technologies can be used in a given society, for the political and social values that a technology expresses or enacts may conflict with the dominant social values. Short-handled hoes are unlikely to gain acceptance among anyone but slave owners. Historians and sociologists of science have pointed out that the technologies of experimental method could not gain widespread acceptability in a slave culture. Experimental method requires a trained intellect as well as the willingness to "get one's hands dirty"; but slave cultures forbid education to slaves and manual labor to aristocrats.[26]

There is a third important relation between science and technology: scientific problematics are often (some would say always) responses to social needs that have been defined as technological ones. For example, scientists are funded to produce information about the reproductive system that will permit the

---

[25]Naomi Schema discusses this problem in her "Commentary on Sandra Harding's 'The Method Question,'" American Philosophical Association Newsletter on Feminism and Philosophy Vol. 88:3 (Spring 1989), p. 40-44.

[26]See Zilsel, Op. Cit.

development of cheap and efficient contraceptives. The development of contraceptives was a technological solution to what was defined by Western elites as the problem of overpopulation among ethnic and racial minorities in the First World and indigenous Third World peoples. However, from the perspectives of those peoples lives, there are at least equally reasonable ways to define what "the problem" is. Instead of overpopulation, why not talk about the appropriation of Third World resources by the First World that makes it impossible for the Third World to support its own populations? Why not say that the problem is the lack of education for Third World women--the variable that is said to be the most highly related to high fertility?[27] After all, just one member of a wealthy U.S. family uses far more of the world's natural resources in his or her daily life than do whole communities of Ethiopians. Would it not be more objective to say that First World overpopulation and greed are primarily responsible for what Westerners choose to call Third World overpopulation? To take another example, research to develop higher yield varieties of grains is said to make Third World farmers more productive. But given the political and economic relations between the First and Third Worlds, what it actually does is to increase the supply of crops for export to the First World, leaving Third World peoples even hungrier than they were before they were the beneficiaries of technological "development." The problem could have been defined as why the First World should profit even further at Third World expense, or why the First World squanders its resources such that it needs to import food from far poorer societies.

This argument distinguishes scientists' intentions from the functions of their work. The point is not that scientists <u>intend</u> to conduct technology-driven inquiry, or to promote the politics that the production of their information requires or makes possible; most do not. Instead, this is an argument about how scientists' research functions within the contemporary social order. This kind of argument is difficult for many people to appreciate, since elites, and especially scientists, are taught to think of the results of science as the consequence of the individual effort of scientists (and teams of them) to find descriptions of the regularities of nature and their underlying causal tendencies that are less false than the prevailing ones. The behaviors of women and members of marginalized races and classes may be a function of their biological or social characteristics, but not the behavior of elites. Elite behavior is the consequence of individual choices and the exercise of will. The contrary argument here depends upon recognition that elite behavior, too, is distinctively shaped by social agendas.

Is there any "pure science" left after we see all of these ways in which science and technology are interrelated? Some would say "yes," that at least one can see in such projects as the search for the basic constituents of the universe scientific research that is not technology driven. However, one can recollect that this research uses technologies that themselves have social implications. (Who is being educated to use them? What kinds of social status accrue to people who get to use these technologies?) Moreover, is it not also reasonable to think that this apparently pure research is justified on the grounds that it is <u>likely</u> to produce

---

[27]See Maria Mies, <u>Patriarchy and Accumulation on a World Scale: Women in the International Division of Labor</u> (Atlantic Highlands: Zed Books, 1986) for a discussion of why capitalist patriarchy cannot permit Third World women to "freely" reproduce themselves and insists that First World women do so.

technologically useful information? Furthermore, the cost of producing this apparently useless information is justifiable to science policy makers on the additional grounds that a small amount of this research has a halo effect on the rest of science. This 5% of "pure research" provides a camouflage for the other 95% that is so obviously technology-driven. If that is its function, how is it pure?[28]

Finally, we might see the insistence on the argument for "pure science" as expressive of a deep irrationality about our culture. In a world where so many go hungry, where cities are in decay, and countrysides have been devastated, where many need medical assistance they cannot afford, where the literacy gap increases between the haves and the have nots--where, in short, access to just a few more resources could have such large effects on the lives of so many--in this world, why should we support scientific activity defined as "pure" precisely because it promises no socially usable results? The support of "pure science" might more reasonably be seen as a make work welfare program for the middle classes, and in the service of elites. It is not that science is responsible for all of these bad characteristics of contemporary social life, but that if it does not develop effective means for identifying the causes and consequences of its own beliefs and practices, it remains complicitous with the production of these social ills.

Here again, in the insistence that the technologies and applications of the sciences are no part of "science proper," one can locate a false belief that we should give up once and for all. It is no accident that sciences that adopt this belief end up disproportionately disadvantaging those, such as women, that elites define as "other."

5. "Scientists can provide the most knowledgeable and authoritative explanations of their own activities. So sociologists and philosophers (including feminists) should refrain from making comments about fields in which they are not experts." To many people, it seems obvious that only physicists can really understand the history and practice of physics; only biologists the reasons why some hypotheses were preferred to others in the history of biology. To hold this view is to hold not the obvious truth that physics should be done by people trained in physics, but the quite different belief that the "science of the natural sciences" is best created by natural scientists--of physics by physicists, of chemistry by chemists, etc. However, if this were so, the sciences would be the only human activity where science recommended that the "indigenous peoples" should be given the final word about what constitutes a maximally adequate causal explanation of their lives and works. It would amount to the same thing to say that there cannot be a science of science; that science alone must be exempted from the claim that all human activity and its products--including the content and form of beliefs--can be explained causally. If we accepted this view, the sciences alone could not be explained in ways that go beyond, or contradict, the understandings its practitioners can produce.

There are at least five reasons why natural scientists are not the best people to provide causal explanations of their own activities. (Most of these claims could be adjusted to apply to practitioners in every other discipline.) In the

---

[28]See Forman's analysis of the loss of purity in 20th Century physics Op. Cit., and Restivo's discussion of this issue in the course of his argument that modern science is today more reasonably considered as a social problem than as a contributor to progress Op. Cit.

first place, a science of science will try to locate origins of everyday scientific activity and belief that are not visible from the location of that activity. In some pre-modern societies, social relations are simple enough that they can be seen in virtually their entirety from the perspective of everyday life. But in modern societies, social relations are complex. In contrast to the case in simple societies, it is impossible to understand how the government, the economy, or the family actually works in complex societies on the basis of our everyday interactions with and in those institutions.[29] For example, the causes of everyday family life are located far away, in the economy, government policy, Supreme Court decisions, child-rearing practices, religious beliefs and other aspects of social relations. Similarly, important causes of scientists' everyday activities and experiences are to be found far distant from the laboratory or field site. They, too, are to be found in the economy, government policy, Supreme Court decisions, child-rearing practices, religious beliefs, and other social relations. A science of science must generate descriptions and explanations of scientific phenomena that start off not only in the labs, but also far away from where scientists and their expertise are located.[30]

In the second place, that "far away" where science begins is temporal as well as spatial. Many patterns in the behaviors of individuals and social institutions are not visible from any single local historical perspective that any individual, or any group such as scientists, might happen to have. They are detectable only if one looks systematically over large sweeps of history. At any present moment, there appear only confusing and small tendencies in various directions. Patterns in these tendencies appear and accumulate power only over decades or even centuries. Distinctive ways of explaining history will be useful in understanding the causes of everyday life in science. Of course, explaining individual events or processes as parts of larger patterns is one way of describing exactly what natural scientists do. The point here is that the history and practices of science themselves can be usefully subjected to such scientific explanations.

But the problem here goes still deeper. Scientists' activity as scientists is exactly the wrong kind of activity from which to be able to detect many interesting causal features of science. For one thing, simply by virtue of choosing to continue to carry out the routine practices of this institution, they undermine the possibility of their providing the kind of critical perspective on those practices that "outsiders" could provide. (I do not say that they cannot provide such analyses--a few practicing scientists in every field have done so.) The same is true of every human activity (including doing philosophy or writing a book!). However, there is a more important reason why scientific activity is the wrong activity for producing causal accounts of science. At least since World War II,

---

[29]Dorothy Smith has made this point repeatedly. See The Everyday World as Problematic: A Feminist Sociology. (Boston: Northeastern University Press, 1987).

[30]This problem is neither resolved nor even acknowledged in the work of the "strong programme" theorists in the sociology of knowledge. See citations in note 12, and Bruno Latour and Steve Woolgar, Laboratory Life: The Social Construction of Scientific Facts (Beverly Hills: Sage, 1979).

doing science has been part of the apparatus of ruling.[31] It generates capital in the form of information, ideas, and technologies that are used to administer, manage and control the physical world and social relations. When human activity is divided in hierarchical ways, those who engage in "ruling class" activity can have only a partial and distorted understanding of nature and social relations.[32] For this reason, laboratory life is especially the wrong activity from which to begin to try to describe and explain the causal relations of administering, managing and controlling the physical world and social relations. Even Kuhn hints at this truth when he points to the false stories about Nobel prizes and glorious careers in science that scientists generate in order to recruit young folks into the arduous training and routine work necessary to science.[33]

In the fourth place, in modern Western cultures, middle-class white men tend more than other groups to believe in the ability of their individual minds to mirror nature, their faculties of judgment to make rational choices, and the power of their wills to bring about their choices, as was mentioned earlier. In terms of the qualities that make them "good scientists," natural scientists are the last people to suppose it desirable to examine the limits of their minds to mirror nature, or to make rational scientific choices, and of their wills to bring about their choices. One could say that they are psychologically the wrong people to provide causal accounts of science. To ask them to try to provide fully causal accounts of their own activity is to ask them to identify the kinds of irrationalities in their own behaviors on which Freud and Marx focussed--not to mention the gender and race "irrationalities" identified by later critics.

Finally, natural scientists have the wrong set of professional skills for the project of providing causal accounts of science. What is needed are people trained in critical social theory; that is, in locating the social contexts-- psychological, historical, sociological, political, economic--that give meaning and power to historical actors, their ideas and their audiences. Natural scientists are trained in context stripping, while the science of science, like other social sciences, requires training in context seeking.

Our ability to understand and explain science would be enhanced if we eliminated the extraneous belief that scientists in general are the best people to describe and explain their activities. To say this is not to say that scientists should be eliminated from the group who can provide illuminating accounts of how science works. Scientists, like anyone else, can use causal accounts of science to generate valuable explanations. But they, like anyone else, must learn how to think about and observe sciences and their technologies in ways for which

---

[31]See Forman 1987 Op. Cit; Hillary Roe and Steven Rose, "The Incorporation of Science," in Ideology of/in the Natural Sciences, ed. Hillary Rose and Steven Rose (Cambridge: Schenkman, 1979). Some historians argue that Newton began this practice; see Van den Daele Op. Cit.1

[32]For discussion of this kind of theory of knowledge, see such feminist standpoint theorists as Smith Op. Cit., Nancy Hartsock, "The Feminist Standpoint: Developing the Ground for a Specifically Feminist Historical Materialism," in Harding and Hintikka, eds., Op. Cit., and Hilary Rose, "Hand, Brain and Heart: A Feminist Epistemology for the Natural Sciences), in Harding and O'Barr eds. Op. Cit.

[33]Op. Cit. 1970.

present-day scientific training does not prepare them. They must become critical social scientists to learn how to reflect critically on intuitive, everyday beliefs about methods and nature that further reflection shows are false. For this reason it could be illuminating to think of the natural sciences as inside, part of, social science.

6. "Physics is the best model for the natural sciences. So feminist social science analyses can have nothing to offer the natural sciences." Now we can consider the false belief that produces the title for this essay. It is still common to regard the natural sciences, and especially physics, as the ideal model for all inquiry. Of course there is a long history of dispute over whether models of research and explanation originating in the study of inanimate nature are the most useful for studying social beings.[34] We need not enter that discussion here because I intend to challenge an assumption made by both sides to that dispute, namely, that the way physics has been taught and practiced--the accepted "logic" of its research processes and forms of explanation--is the best that it could be: that "physics" is a good model for physics. Both the "naturalists" and the "intentionalists," as the two parties have been named in the philosophy of social science dispute, assume that physics provides a perfectly fine model of inquiry and explanation for the natural sciences. That is not controversial to either group. Though--my point here--it should be. So my argument will be not that physics provides a poor model for social inquiry, but the stronger argument that the paradigm of physics research and explanation, as it is understood by scientists and most other people is a poor model for physics itself.

We can appreciate the historical reasons why the physics of the 17th and subsequent centuries was so highly valued as a model for all scientific inquiry. In the 20th century, the unity of science thesis of the Vienna Circle provided the modern justification for prescribing a hierarchy of the sciences with physics at the top. Ironically, the analysis in this essay can be understood to agree that the sciences should be unified.[35] However, from the perspective of the analysis here, the hierarchy should be "stood on its head." On scientific grounds, as well as for moral and political reasons, those social sciences that are the most deeply critical and the most comprehensively context-seeking can provide the best models for all scientific inquiry, including physics.

It is not helpful from a scientific perspective to have as the model of science research projects in which controversy about basic principles is absent-- the criterion Thomas Kuhn used to identify research projects that had reached the truly scientific stage (as we noted earlier). The problem with Kuhn's criterion is that in sciences that are important to dominant groups in socially stratified societies, lack of controversy about fundamentals is not a reliable (or even plausible) indicator of the absence of social, economic, and political values in a research process or its results. This is such a society, and physics is such a science. Perfect agreement about basic principles and methods of inquiry can be and has often been reached by scientific "fast guns for hire" employed by the most egregious sexists, imperialists and profiteers. Even more distressing is the history

---

[34]See the summary of it in Brian Fay and Donald Moon, "What Would an Adequate Philosophy of Social Science Look Like?" Philosophy of Social Science 7 (1977).

[35]I discussed this in Harding 1986 Op. Cit.

of well-intentioned research by the most distinguished of scientists that was inadvertently highly constrained by the sexist, racist, imperialist and bourgeois ethos of its period.[36] Instead, the model for good science should be research programs that have been explicitly directed by liberatory political goals--by interests identifying and eliminating the partialities and distortions in our understandings of nature and social relations that have been created by socially coercive projects. It does not insure good empirical results to select scientific problematics, concepts, hypotheses, research designs, etc. with these goals in mind; democratic sciences must be able to distinguish between how people want the world to be and how it is. But better science is likely to result if all of the causes of scientific conclusions are thought to be equally reasonable objects of scientific analysis. Since sexism, racism, imperialism and bourgeois beliefs have been among the most powerful influences on the production of false scientific belief, critical examination of these causes, too, of the "results of research" should be considered to be inside the natural sciences. Or, we could say that the natural sciences should be considered to be embedded in the social sciences since everything scientists do or think is part of the social world.

Objections and Responses. Such a proposal will seem bizarre to thinkers who are more comfortable with the scientific and epistemological authoritarianism embedded in models of "value-neutral" research that dominate in the natural sciences. Let me clarify this proposal by responding to some predictable criticisms of it. I shall be repeating in different terms the arguments above, but a small amount of repetition can be helpful.

Here is one: "who is to decide what is liberatory? What's liberatory for you may not be so for me." It is true that people will have to negotiate with each other through personal and political processes about just whose perspective should prevail when "facts" are being contested. If those processes are not now sufficiently democratic, then we have to take (democratic) steps to make them better. But this problem of "whose perspective?" is not solved by hiding that decision process behind claims to value-neutrality. Many scientists do not really believe--and some actively protest--the dominant scientific ideology. But the myth of experts and their authority is, nevertheless, the one used to recruit students into science educations and to keep the sciences linked as firmly as possible to the projects of the dominant groups in the West. Thus many people who are most comfortable with hierarchical decision making and have little experience in negotiating social arrangements except among white, Western, economically privileged, men like themselves will find it difficult to participate effectively in these negotiations. (But it is never too late to learn new skills!)

Another objection: "Discussions of the appropriate goals of science should indeed occur, and of course the needs of minorities, women and the poor should be considered. However there is no good reason to think of these discussions as part of science itself. These are discussions more appropriately conducted in political arenas rather than in laboratories or other locations where scientific research is done." However, moral and political loyalties have counted as part of the evidence for the best as well as the worst hypotheses in the natural

---

[36]See Gould Op. Cit. and Forman 1987 Op. Cit.

sciences.[37] The problem is not primarily the commitments of individual scientists that differ from the commitments of other scientists, for those differences are relatively easy to identify and eliminate from research processes through existing norms of inquiry. The problem, instead, is those values, interests and commitments that are close to culture-wide within scientific cultures or cultural elites, for these cannot even be identified by the methods of the natural sciences. If all of the evidence for scientific belief is to be critically examined, so, too, must these social commitments that function as evidence.

Objection: "But I thought it was exactly widespread social beliefs that the individual critical observation and reasoning of the sciences was supposed to correct. It is individuals in the history of the sciences who have formulated hypotheses, observed nature, and interpreted the results of research. The Great Man history of science may not be the whole history, but it is a distinguished and central part of it. You are simply proposing that science be entirely subjected to mass thought and thus to the irrationality of politics." However, Western scientific thought, no less than the thought of other cultures, has distinctive cultural patterns. I always see through my community's eyes, and begin thought with its assumptions. Or, in other words, my society can "observe" the world only through my eyes (and others'), and can only think beginning with my assumptions (and others'). In an important sense, my eyes are not my own, nor are even my most private thoughts entirely private. My eyes and my thoughts belong to my historical period--and to the particular class, race, gender, and cultural commitments that I do not question. It takes a reorganization of the scientific community, and a rethinking of its goals and methods, to make visible the social characteristics of the purportedly invisible authors of claims in the natural and social sciences. We need to be able to see how gender, race and class interests shape the projects of laboratory life and of the manufacture of scientific knowledge. This, too, is a scientific project, and one that can usefully be regarded as part of the natural science.

Objection: "Aren't you arguing that we should substitute subjectivist and relativist stances for objectivity in the sciences?" On the contrary, any research that is conceptualized as maximally value-free on the grounds that--among other things--it does not critically examine the social causes and dimensions of "good" as well as "bad" scientific belief is, I have been arguing, disabled in its attempts to produce objective understandings of nature and social life. It is unable to critically scrutinize one of the significant causes of widespread acceptance of scientific hypotheses. We need a notion of "strong objectivity," rather than only the "weak objectivity" on which conventionalists insist. Nature causes scientific hypotheses to gain good empirical confirmation, but so, too, does the "fit" of these hypotheses problematics, concepts and interpretations with prevailing cultural interests and values. It is a maximally objective understanding of science's location in the contemporary international social order that is the goal here. This is far from a call for relativism. Instead it is a call for the maximization of criticism of superstition, custom and received belief for which the critical, skeptical attitude of science is supposed to be an important instrument. Ironically, we can have a science of morals and politics not by

---

[37]This is another way to put the point made by Van den Daele Op. Cit., Merchant Op. Cit., Keller Op. Cit., Forman Op. Cit. and others.

imitating the natural sciences in research projects in these fields, but only if we put critical discussion of morals and politics at the heart of our sciences.

Objection: "Isn't this argument really against science? Aren't you 'down on physics'?" No doubt many will think so. But this argument has a different target. It is against a certain kind of narrow, no longer useful interpretation of why it is that physics has contributed so greatly to the growth of scientific knowledge in the West. "Sciences for the people" (in Galileo's phrase), not for elites, should be the only way support for science can be justified in a society committed to democracy. There are plenty of useful projects for such sciences, but they do not include research that provides resources for militarism, or ecological disaster, or that continues to move resources away from the underprivileged and toward the already overprivileged.

Conclusion. There is plenty of science still to be done once physics is invited and permitted to step down and take its place as one human social activity among many others. What kinds of knowledge about the empirical world do we need in order to live at all, and to live more reasonably with each other on this planet from this moment on? Who should make up the "we" who answers this question?

QUESTION

HESSE: I'd like to thank Sandra Harding for her very clear and thorough exposition of a postmodernist position, and I found myself in agreement with almost all her substantial criticisms of the image of modern science. The trouble is that I am not a feminist. I don't think I'm an imperialist, a sexist, a socialist, or a racist, or any of those labels, but I have, I think, a much more radical interpretation of the critique of objectivity than she does. When she reached her last paragraph, in which she spoke about a transformed objectivity, I was just about to write down the question--How is the acceptance of the falsity of the various beliefs going to help us to hold a more rigorous objectivity? The thing that worries me is this, that we're looking for an objectivity which is less biased and distorted by social contexts and origins, while at the same time being itself not value free. So we're looking for something which is less biased and yet not value free. Now how can we have something which is itself value laden and yet is objective without changing the whole sense of what objective means?

HARDING: First of all I want to agree with you that I think it's a very conservative notion of objectivity that I'm in fact proposing here. Not everybody might agree with you and me, but in fact I am trying to suggest that there are important aspects of the traditional notion of objectivity which need not be challenged in order to accomplish the goals that I have in mind. The reason why the conservatism is valuable is that it enables the critique of the ecologists, the feminists, the postcolonialists, to be heard within the conventional discourse in a way that is more radical.

Now to your suggestion that there is a paradox in the notion that there is an objectivity which is less biased, which nevertheless is still value laden. Let me talk for the moment about feminism. When feminist critics argue that a particular theory in biology or sociology is sexist, they are simultaneously commited to a set of values, namely that sexist theories should not be around. So they are value committed in that way. They are not proposing that we substitute a loyalty to femininity for the loyalty to masculinity that has been undetected in the offending theory. So they are proposing that by seeking out how gender shapes our beliefs about the world, that commitment results in more beliefs about the world, that commitment results in a more objective set of claims about the world. Galileo was creating a science which was to be more objective than medieval science. This is similar. The reason why I think this appears paradoxical is that people hold a peculiar notion about the ethics of modern science. In fact, it is built on a democratic ethic. It says all observers are equal, knowledge is public, and everyone can comment on it. It builds in a particular set of values while simultaneously claiming to produce knowledge claims that are not partial to any social group.

The Death of Science!?

SHELDON LEE GLASHOW

I am honored to have been called to Gustavus Adolphus College to deliver a talk at this twenty-fifth officially authorized Nobel Symposium. Past meetings in this prestigious series dealt with specific issues in Science or Economics. The subject material of this Symposium, to put it politely rather than scatologically, is softer. This is not the kind of talk I like to give--I prefer discussing the substance of Science or the teaching of Science. When Science itself becomes the issue, political, sociological, or philosophical, I am well off my normal turf. What I shall say today would be obvious, trite and trivial to my colleagues and my students, and hopefully it will be to you as well: that Science is nearer its beginning than its end; that Science offers the last great hope for a humane and lasting society; that the pursuit of Scientific understanding can be itself a primary human goal; that Science is the unique truly international forum and has been for at least five centuries. Tycho Brahe, Copernicus, Kepler, Galileo and Newton: five men from five nations taught us our place in the universe. And today, at one of the very few international organizations that really works, fifteen European nations operate the world's premiere particle physics laboratory, CERN. Their scientists, and those from a dozen other nations (including our own, China and Russia), actively pursue the purest (and the most pointless, for its potential technological relevance) scientific research.

The scientific enterprise in this country is far from ending, but it is certainly slowing down. As fewer bright children choose Science as a career, our scientific, technological, and pedagogical needs are more and more being met by foreign imports who are far better trained than our domestic products. The reasons underlying this severe national problem have little to do with Science *per se*, but with our often inept teaching, ill-motivated and poorly-guided youth, and the short-sightedness of our government and many of our industries.

The pursuit of my own discipline of elementary particle physics is threatened from an entirely different direction: from its very success. Our so-called 'Standard Model' is clear about what questions it can and cannot answer. It does not say why there are muons, or what the quark masses must be: questions, we feel, that a satisfactory theory must respond to. Yet, those questions that the standard model does address, it answers very well. The last decade of research in high-energy physics has produced very many confirmations of the theory, but has revealed not the slightest flaw, not the tiniest discrepancy from the expectations of the standard model. We have no experimental hint or clue that could guide us to build a more ambitious theory. However, Nature's road has often seemed to be impassable but we have always overcome. We particle physicists are perhaps unique among scientists: we strive to undermine today's theory so that tomorrow's may be even better.

But, these problems are not what this Symposium is about. The Big Questions here are: Is Science simply a grown-up version of a boys-only game of 'Let's Pretend'? And, can a new kinder and gentler Science be invented with

Force replaced by Love, and Power by Tenderness? "No, and no again," is a brief summary of the rest of my talk.

The End of Science means different things to different souls. Some scientists worry that we have reached a plateau of understanding from which further progress will be difficult and unlikely: because, in view of the great success of our theories, the future can be more than commentary on, or elaboration of, past triumphs; or because, in view of the enormous expense of our envisaged endeavors--the Advanced X-ray Satellite, the Supercollider (or Gippertron) and the Genome Project being three examples--we can no longer afford the luxury of such pursuits. Some non-scientists think that Science is pernicious, that it is responsible for the pollution, contamination and despoliation of our air, our water and our earth. Tried *in absentia* and found guilty, the scientific enterprise should be terminated. My distinguished colleague Sandra Harding, on yet another hand, finds that Science as an objective search for knowledge must be *"reinvented"* from scratch. Science, as it is and always has been, is *"inextricably connected with specifically masculine needs and desires"* and it cannot serve to *"make sense of women's social experience."* In her view, it seems to follow that *"there is nothing morally and politically worth redeeming in the scientific world view."*

I subscribe to none of these views. I find them absurd. I believe that Science has contributed immensely to our health, welfare and fulfillment, but that we have yet to realize its full potential. I believe that we have made enormous strides in understanding the mysteries of Nature, Life and Mind, but that even grander challenges remain to be met by future generations of scientists. Only by the wise application of Science can we learn to live in peace, comfort and harmony as the crew of Spaceship Earth. Only by the understanding of Science can we come to terms with our own mortality, as well as that of our species and our planet.

In this morbid connection, recall that God, Art, History, and even Communism have already left this mortal coil. No discipline is immune. Washington University, in Saint Louis, is liquidating its Sociology Department and today this august assembly examines the entirely inane proposition that Science, too, has succumbed. Whatever does it all mean, and why on earth did I agree to participate in a farce? Let me begin with a modified and annotated extract from my irresistible and unforgettable letter of invitation to this Symposium:

Science as a unified, universal, objective endeavor is currently being questioned. By whom? The letter doesn't say, so I must embody an imaginary and genderless Sidney to act as our courageous, if misguided, inquisitor.

Sidney regards Science as a more subjective and relativistic project. No comment on the incomplete comparative--substance not syntax is my beat. But I believe that the laws of physical Science are objective: they are among the few things agreed upon by all of the family of man. I will not accept the media metaphor: that scientists fashion the underlying laws of Nature much as a reporter fabricates the news. Experiments are done and can be repeated. The results either fit with or falsify a theory. (A theory that simply can accommodate itself to any data is simply not a theory at all.) Social scientists, perhaps dissatisfied with the imprecision of their disciplines, often misuse our notions of quantum mechanics and relativity to suggest that physics is a social science as well: misery seeking company. Quantum mechanics does say that every observation is invasive: the experiment affects the system being measured. But quantum mechanics does not say that the social, religious, political, or even scientific views of the experimenter can influence the result of the experiment. Relativity says that events happening simultaneously for one observer need not do so for another. Relativity does not say that the truth of a fundamental law may vary from one culture to another. Neither relativity nor quantum mechanics casts any doubt on the existence of an objective and external set of knowable rules governing all observable natural phenomena.

He/she believes that Science is currently re-examining itself as the product of paradigmatic foci, ideological struggles and the basic instruments of power. No scientist I know is undergoing such an agonizing and polysyllabic reappraisal. Curious, is it not, that the sternest critics of Science are so often those who know it least?

Sidney doubts that Science reflects extra-historical, external, and universal truths. Science, Sidney feels, is social, temporal and local, and it evolves like a species or a society. Since Science is what we know about Nature, and since what we know is a function of time, place, and social accident, there is simply no way of speaking about something real behind Science. Science is merely an account of the imaginations of the observers of Nature. Moreover, Sidney sees Science as culturally-influenced and even repressive, and hardly as a fortress of objectivity. That is the sense in which Science as Science is presumed to be dead.

Them's fighting words, so you see why I had to come to this meeting. Thousands of impressionable students, all of whom should study some Science but probably will not, are to be exposed to those who would bury Science not praise it. I shall struggle as best I can with what I have now been told is a Grave Epistemological Issue. Forgive me if I thrash about a bit--it's not easy to beat a dead oxymoron.

There have always been those who would deny the reality of physical constructs that are not a familiar part of everyday life. When Galileo spied the moons of Jupiter, a seemingly incontrovertible discovery, he was assured that:

It is necessary to say that the poetic imagination comes in two variables: There are those who invent fables, and those who may be disposed to believe them.

So, in effect, says the letter that I have just so gently ridiculed.

The lay community almost always confused the notion of Science with that of its close cousin Technology, which is certainly socially influenced and culture dependent. What good would penicillin be to an alien life form based on compounds of silicon, or birth control pills to parthenogenetic extraterrestrials? Would Burroughs-Wellcome have developed AZT had there been no bath houses in San Francisco and intravenous abusers in New York? Why should firemen care about fortran or Eskimos buy air conditioners? All these things are products of Technology not of Science.

Technology is the designer-fruit of Science. Society must decide on its nature and must wisely choose its applications. Science enables Technology, which opens up enormous and rapidly growing ranges of human opportunities. Life, liberty, and the pursuit of happiness are accessible, in principle, to all of us:

so also are death, slavery and the endurance of misery. Society must determine its means and its goals. The choices that must be made are not scientific, although each decision may have its technical aspects, and science may tell us what can be done and how to go about doing it. We all agree that Technology is influenced by political, societal, and commercial forces, but don't blame the drivel of American television programming on us scientists, nor our firm but flavorless tomatoes, nor our preference for Toyotas to Fords, nor the AIDS epidemic, nor the ozone hole, nor the threat of nuclear winter, nor the lack of female mathematicians and of black astrophysicists. These are real problems, but they are more societal than scientific.

Science impels Technology. Quantum optics led us to lasers with which to repair our damaged eyes. Nuclear physics promises us, and provides the French, a source of power free from oil spills, acid rain, black lung disease, strip mining, and a greenhouse effect. Chemists devised new materials which have become household necessities: like teflon and nylon, superglue and silly putty. Biologists made possible the abolition of such scourges as smallpox, polio and pimples. Condensed matter physics--once called Solid State, a phrase now preempted as a trademark--gave us transistors and thereby pocket calculators, video games and Walkmen. Yet, Technology is not merely the useful offspring of Science. The linkage is more incestuous because the parent-child relationship is often reversed.

Science *is* culture dependent because Technology fuels the progress of Science, and Technology is hostage to social perversion. Arabic astronomy, chemistry and mathematics once were unrivalled.. Early in this millennium, *Taqlid*, the doctrine that no truth exists beyond that revealed in the Korean, was instituted in the Islamic world. Scholars were banished. The only remaining trace of their Science is in our language, in words from Alkali to Zenith. The Chinese too had their fling with Science. They invented gunpowder, the compass, and were the greatest navigators on earth until they decided, in the 15th century, that nothing beyond their Celestial Empire was worthy of discovery. They burnt their great ships just before Columbus set forth with his tiny flotilla to discover a New World.

Science is culturally driven because its progress depends upon technological innovation. A Dutch optician happened upon the invention of the telescope. Galileo, hearing about the device, soon built his own, got his salary doubled, and discovered mountains on the Moon, phases of Venus, spots on the Sun, and satellites of Jupiters: four genuine examples of objective reality. Only the Church thought otherwise, and sadly, perhaps a few of us. Another Dutch invention, the microscope, opened up the wee frontier of microbes and led us inexorably to develop a science of life. Germs are seen and killed, not imagined and unimagined. The compass was more than a child's toy and a navigational tool: it was the key to H. C. Oersted's discovery of the intimate connection between electricity and magnetism. Newly-devised electric batteries, in the hands of Sir Humphrey Davy and Michael Faraday, generated major breakthroughs both in chemistry and in physics. Bunsen's burner revealed the characteristic spectra of the chemical elements, the nuclear reactor revealed the neutrino, and the bubble chamber was our window to the world of elementary particles. Technology begets Science, which generates new Technology in its turn.

American taxpayers or their elected representatives choose to spend hundreds of millions of dollars each year on the abstract disciplines of particle physics and cosmology with little thought and no demand for an immediate, or even an eventual, pay-off. However thankful I am for their generosity, I am certain that charmed quarks, W-bosons, pulsars, and quasars are not a mere phantasmagoria of the collective scientific imagination, nor were they imposed upon us by the funding agencies. These hidden wonders of the natural world are there to be seen like atoms or like the recently found rings around Neptune: by Americans, by Russians, by Japanese, by Ugandans, by men and by women.

Our remarkable insights and triumphant theories are socially affected only in the sense that society pays the bills. Quarks are a reflection of our culture only in the sense that we were smart enough to figure out that we are made out of them. So are any subjectively imagined culture-laden intelligent asexual ameboids, somewhere in a galaxy far far away. Their understanding of the physical universe must be much like ours, although they might differ from us on what are the Gravest Epistemological Issues.

I shall focus on the thesis that Science is totally objective and is in no sense related to or dependent upon Humankind's particular circumstance. To be as incisive as possible, let me include within Science only such disciplines as mathematics, physics, chemistry, astronomy and cosmology. I do not say that biology, psychology, paleontology, and terrestrial geology are not Sciences. I exclude them, for the sake of this discussion, because they are somewhat tied up with our planet and ourselves. Life on earth is built up out of very specific organic compounds like RNA, DNA, and amino acids. For all we know, it could have been, and elsewhere well may be, quite otherwise. Although the study of human physiology *is* objective and not culture-ridden, it is also species and site specific: to us and to the earth. Let us avoid the complications inherent in such questions as whether sentient life on other worlds exists but follows other patterns, or whether human life begins at conception. The very nature of our planet is affected by its inhabitants, so that the study of earth and its creatures great and small *is* of immediate societal impact. Plants, over the course of a billion years, made our atmosphere what it is today, but *we* shall determine what it will become in a century. How many species do we carelessly obliterate? How many rain forests have we cut down, elephants mutilated, pristine streams polluted? The excluded Sciences, moreover, are those that more than any others have been subject to pernicious, corrupting and overbearing societal forces such as Nazi eugenics, Freudian psychology pretending at Science, and Lysenkoism. No comparable campaigns have been launched against the transcendental nature of π, the atomic hypothesis or the special theory of relativity. Indeed, those German scientists who remained within the Third Reich, knowing the truth, argued that the so-called Jewish theory of relativity should be accepted by the Nazis because it surely would have been discovered by an Aryan had Einstein never lived.

Science, in the truncated sense I intend, depends upon two unspoken but fundamental assertions. Neither assumption can be proven, and neither is even obviously true. Our faith in these principles is partly religious and partly pragmatic. We believe that the world is knowable: that there are simple rules governing the behavior of matter and the evolution of the universe. We affirm that there are eternal, objective, extra-historical, socially-neutral, external, and universal truths. The assemblage of these truths is what we call Science and the proof of our assertion lies in the pudding of its success. The second principle amplifies the word universal: we believe that the laws of Science are the same everywhere and everywhen, that our time and place in the universe is neither chosen nor distinguished nor special. Natural laws can be discovered that are universal, invariant, inviolate, genderless, and verifiable. They may be found by men or by women or by mixed collaborations: they are not the *œdipal* preoccupation of males. Any intelligent alien, anywhere, sooner or later, would have come upon the same logical system as we have to explain the structure of protons and the nature of supernovæ.

While our faith may be irrational, its success is undeniable. Percy Bridgman observed that those scientists who believe in the underlying simplicity of Nature are those who are successful in their quests. Nuclear physics, deduced by earthlings from a remote corner of an unexceptional galaxy, explains the behavior of the stars we see throughout the universe. Occasionally a star 'goes supernova,' suddenly becoming a billion times brighter than it had been. The last supernova in our Galaxy took place in 1604. It was visible at mid-day and it was bright enough to read by at night. Astrophysicists understand what a supernova is, and they estimate that 99.99% of the energy it releases should emerge as an intense but invisible flash of ghostly neutrinos. Thus, the titanic explosion we see as starlight is but the palest shadow of the death of a giant star. Two years ago, a new star appeared in the Southern Hemisphere as light from the first relatively nearby supernova in almost four hundred years arrived at our planet. Gigantic detectors situated deep under the ground in Ohio and in Japan looked for the predicted neutrinos, and they were there! Scientists had correctly analyzed an event that took place hundreds of thousands of years ago and a billion miles away. Their prediction, now established by observations, was based upon the universal laws revealed by experiments performed by the men and women of planet earth.

Is there not a sacred covenant that we must understand as best we can the world we are born to? Is not our comprehension of nature as proud an achievement as our Art, our Music, our Literature? Can one doubt that Culture and Science are indivisible? Surely there is no need to sing Science's praises nor to participate in this premature wake for what is in fact a vibrant youth. The glorious images that Voyager has given us of the outer planets and their moons and rings are accessible to every couch potato. They are no fancy of NASA's minuscule imagination. The millions of Z bosons which will soon be produced and studied at the gigantic new European accelerator LEP are real things of nature. The protogalaxy just serendipitously (and supposedly) observed by a man and a woman working together at the immense Arecibo radio telescope is no human whim: it is what the Milky Way may have been like billions of years ago. And, Margaret Geller taught us just last year that the distribution of galaxies in the universe resembles, in her words, *the suds in a kitchen sink*, and another woman's work established the existence of mysterious and massive invisible halos surrounding each galaxy. (Cold fusion, however, an irreproducible effect alleged to have been seen by a few men in a four-letter state, I accept as human foible.) So what is bothering our paradigmatic sociopaths?

All the hoopla about the death of science may have something to do with scientists' rather careless and casual use of the words *theory* and *truth*. Take for example, Boyle's Law, that there be a spring to the air, one of the first precise quantitative laws to be established describing the behavior of macroscopic matter: double the pressure on a gas, and its volume will halve. It is a fundamental law, and it led us by the nose to believe in atoms. But is it true under all conditions? Of course not: squeeze any gas too much, and its atoms will touch one another;

the stuff liquifies and Boyle's Law no longer applies. Even at ordinary conditions, small departures from the law can be measured which result from the finite size of atoms. So Boyle's law is wrong! The same goes for Newton's laws: they are fine for predicting the trajectories of ICBMs or the times of eclipses. But, they make a tiny error in the description of Mercury's orbit, and they fail to describe the bending of starlight by the Sun. So Newton's laws are wrong. So are Maxwell's laws. So are a lot of other so-called laws: and maybe even all of them. So what is Science other than a bunch of old-fashioned broken-down mostly man-made laws. What is true today may be false tomorrow, so how can Science be anything other than a societal construct, like Keynesian economics, Laffer's law, marriage, or the rules of point-count bidding?

The misunderstanding is simply this: Boyle's law and Newton's laws and all those other laws are true after all. It's just that they have a limited range of validity. Within their domains they are true, they are true, and they will remain forever true. Unlike the social sciences, for which there is no absolute standard of truth and in which diametrically opposed views can perpetually coexist, physical Science is a vertical construction and any alteration or addition to its edifice must be compatible with what is already firmly in place. Falling bodies are still uniformly accelerated a la Galileo, although sometimes they may be made to circle the Earth. We still live under Torricelli's ocean of air even though its CO_2 content has dangerously increased. Maxwell's equations are still valid, but they have been quantized so as to explain how light is neither wave nor particle. The new theory must comprehend the old, as quantum theory and relativity reduce to classical physics in the slow and clumsy work-a-day world. Truly, we see so far today because we sit on the shoulders of such giants as Galileo, Newton and Boyle. Their results are recapitulated today in terms of a far more powerful and wide ranging theory, but a theory that accepts their truths and offers many more. Newton's laws are true, but Einstein's are even truer. Newton's accomplishments are as great as ever they were, but they are embedded within a grander scheme encompassing galaxies and their inhabitants as well as atoms and their constituents.

Consider the search for the ultimate constituents of matter and the rules by which they combine, the logical structure underlying all of Science, the discipline now known as elementary particle physics. In the 19th century, some scientists suspected that atoms might be more that a useful mathematical construct or mnemonic with which to codify the empirical rules of chemistry, to model the behavior of gases, and to explain the growth of crystals. They began to believe that atoms really do exist, and so what if you couldn't see them? (After all, you could only see the moons of Jupiter with a telescope, but this was no longer a reason to deny their reality. In any case, today's scanning tunnelling microscope does allow us to single atoms.) The size and mass of an atom was measured, as well as its spectral lines and its chemical properties. Men and women of all races and creeds were forced by the data to conclude that atoms are genuinely real things.

Their existence conceded, it was only a matter of time before scientists would ask of what simpler things be atoms made. There were simply too many different kinds of atoms for them to be elementary. Electrons and nuclei were discovered--not imagined or contrived, mind you, but observed and measured in the laboratory--and Niels Bohr introduced a set of *ad hoc* quantum rules to explain how they could be assembled into atoms, since the classical rules no longer work in the arena of atomic sizes. Bohr's mysterious rules evolved into a consistent theory, quantum mechanics, and scientists went on to unravel many of the secrets of matter perceived: why copper is red, the sky is blue, water is wet and diamonds few. They are still at it, for example, trying to understand how the new and technologically promising high-temperature superconductors work. It's like an impossibly intricate game--the quantum rules of atomic combination are set in stone like those of chess: they are absolutely true in the relevant domain. Scientists must use their experimental ingenuity and their theoretical insights to try to figure it out. So they shall, but the solution, found by mere mortals with mortgages to pay and children to raise, will be another of our verifiable and predictive truths. That these superconductors behave as they do is neither a social accident nor an imaginary account. Science is not sick and it is not dead: it thrives as it answers our deepest questions. The greatest mystery of all, and one that Science cannot possibly solve, is why the universe seems to be comprehensible to us.

Other scientists wanted to learn what a nucleus is. Half a century ago, they found it to be made up of protons and neutrons, things even harder to *see* than atoms. As cosmic rays were studied and large particle accelerators were deployed, lots more particles were discovered: hundreds of them, with names like California license plates. Ultimately we realized that most of these seemingly elementary particles are not elementary at all. They are made up of quarks. The reality of these objects (or constructs) *is* vulnerable to philosophical inquiry, for quarks present an engaging epistemological puzzle. There is no such thing as a piece of string with only one end. Analogously, we believe that the quark cannot exist as other than a part of an observable particle. The quark itself cannot be isolated. Moreover, quarks are held together by the effects of *gluons*, which are themselves not isolable. At this point, the border-line between mathematical invention and physical reality begins to blur.

Perhaps, you say, the quark is truly a mere mathematical construct, a useful artifice with which we may organize knowledge. Even Murray Gell-Mann, who introduced the name and the notion of quarks, once held this view. But our experimental skills and our theoretical understanding have advanced to the point that we can *see* the quark, if you grant us some leeway in the concept of seeing. In reality, we do not really see much of anything. All we do is to sense some of the photons entering our eyes. When we see an apple we merely react to the light it reflects. The apple exists in the dark, to be touched or smelled or eaten or imagined. It is detectable only because it can interact in an observable way with other forms of matter or energy. Concede this, and you have conceded the existence of quarks. We do not see a quark quite as we see an apple because light is too feeble a probe. Nor can we taste or smell or hear or touch it. But we can and we do detect its effects and its interactions, such as the jet of particles which is the spoor of a newly born quark, from which we can determine its properties in detail. We have so far identified five distinct species of quark, and we have measured their masses, their electric charges, *etc*. Quarks are neither more nor less real than apples or atoms.

Social forces can affect the rate of scientific discovery, but not the eventual outcome. Let me give an example in which I played a minor role: the experimental search for what are called *neutral currents*, culminating with their observation, in Europe and in America, in 1973. This discovery was among the first confirmations of the unified theory of weak and electromagnetic forces, a theory which has met every experimental test and is now generally accepted to be valid, as always, within its domain. However, neutral currents could have been found a decade earlier. They weren't found because they weren't looked for. They weren't looked for because very few physicists were aware of the electroweak theory. And, that's because it really wasn't very much of a believable theory until Gerard 't Hooft in 1971 showed that the theory made mathematical sense. Only then did the search for neutral currents begin in earnest. We see that theoretical physicists, through their endeavors, can establish a climate propitious for experimental discovery. But the object to be discovered has got to be there to be found. Proton decay, majorons, axions, and magnetic monopoles are examples of brilliant theoretical inventions that have not been confirmed in the laboratory. No matter the power and the beauty of theoretical argument nor the will of the masses, repeatable objective experiments remain the only basis on which to judge the reality of the creates of the scientific imagination.

We particle physicists, as the unreformed reductionists we often are, have reason to believe that there are at least seventeen apparently fundamental particles in Nature, of which all but three have been directly *seen* in the laboratory. This is a large number of basic building blocks--four was enough for the ancient Greeks-- and it seems to contradict our faith in the simplicity of Nature. Surely our universe should have simpler beginnings.

For this reason, many theoretical physicists today struggle with a strange discipline of mathematical physics called superstring theory. They seem to have created the first consistent quantum theory of gravity, and with some luck, they may one day understand all of the particles we see as twists of a tiny loop of string embedded within a ten-dimensional space-time. We cannot know whether these ideas represent creatures of thought or of reality, whether they will go the way of phlogiston or of quarks. Superstring theory is truly at the speculative frontier. Much the same can be said about contemporary cosmology, with its inflationary scenarios, its wormholes, its baby universes, and the notion of an eternal and self-replicating reality. Here, at the newest and fastest growing roots of particle physics and cosmology, we are admittedly unsure (yet!) of whether or not there is something substantive behind our Science. However, this is a symptom of the health and vitality of science, and is an indication of its imminent end. Uncertainty is what makes these wild disciplines such fun and so much of a challenge.

Every new idea in Science, at its inception, is certainly a function of time and place and social accident. And, most of these new ideas, evolving much like species or cultures, make themselves extinct simply because they do not correspond with external reality. The heap of discarded hypotheses and trashed theories is far larger than the edifice bushed knowledge. Only a very few ideas turn out to survive experimental scrutiny, to be valid within their domains, to be external, objective, and, yes, even true! What joy such an event produces for the lucky scientist by whose imagination a hidden aspect of reality becomes known to all and forever! Pity the social scientist, or the ideologist, or the pseudo-philosopher who may never know anything with such certainty. Long Live Science!

QUESTION

HACKING: Now, I share with Professor Glashow the belief in the existence of most of these things with which we interact. What struck me, what I would like to ask him about in his presentation, was this repeated conception of having the theory and then examining the inert world, and thinking of the theory as answering to the inert world by means of the experimenter, who just sees the quarks, or whatever. It seems to me that what the experimenter does is to create that subject matter to which the theory applies, and we modify our apparatus and our theories in order to bring them into a sort of harmony, a harmony which I believe accounts for that talk of his, about theories being right in their domains. As if they were domains of a passive world. Rather, they are domains of an active world with which we interact. And each of these domains is not, in my opinion, a part of the world that eternally exists, but rather a world of apparatus and of interactions that are ours. I believe that reflection on this made it somewhat more possible to understand the very thing which he said was in principle not understandable, namely how we are able to have such knowledge of the world, which would be really much more baffling if it were knowledge of a totally inert world. This consideration also diminishes for me the certainty Glashow has that any creature, anywhere, following any history, would reach the same conclusions about what we are made of, if they got into the "what we are made of" business at all.

GLASHOW: That's a hard position to understand, but I'll try to respond as best I can. First of all, I'm not really going to answer it. I'm going to object at one tiny level where you say we modify the apparatus, as we surely do, building larger and larger accelerators, and modify the theory, as we surely do. We try to make it better. At the moment, however in my game, we are in this curious little phase, which has lasted for a decade, where we sit around praying, we theorists, for experiments which vitiate the theory we have created. No matter how the experimenters twiddle their dials, and build new equipment, and spend billions of dollars, they simply cannot, much against our will and our desire, find any flaw in our theory. And that's kind of a different situation from the one you are describing, where instead of trying to make nature agree with us, we want nature to disagree with us, but it won't do it! We seem to have a theory that we can't shoot down, but, which does not answer a number of very profound and important questions.

Now that's one point. The second point is we seem to differ at an extremely profound point, at the initial point of our faith, wherein, I believe that my amoeboids in another part of the universe would have the same fundamental understanding, if they were so inclined to be interested in such questions of the structure of the proton, they too would believe that it is made of three quarks which come in several varieties, etc., etc. On the other hand, Ian does not have this faith and therefore there is an unbreachable divide between us because we simply do not share that one faith which I cannot justify.

Disunified Sciences

IAN HACKING

The end of science! That would not have to be the end of the road. Perhaps, like infants, civilizations go through a series of stages and we're about to move on past an era in which it was king. August Comte thought that progress was like that. He was the man who, early in the nineteenth century, invented both positivism and sociology. People had a habit of stealing his catchy names. He thought that 'sociology' was such an ugly word that nobody would ever take it away from him, but he was wrong. Noticing that in all the European languages the word 'positive' means nothing but good, he thought that 'positivism' would be a great name for his new philosophy. He was wrong about that too. Positivism is now a Bad Word. Maybe our conference is more about 'The End of Positivism' than 'The End of Science.' But had we called it that, no one would have come.

Comte said that the human race--he meant Western Man--had passed through two stages of intellectual and spiritual development, one theological, one metaphysical. By the 1830s, he taught, we had entered the stage of positive science. Theology and metaphysics had gone or were on the way out, and would stay out so long as alert modern men and women kept them down. He did know that people need weekly spiritual uplift, so he helped found a positive church, whose higher power was Humanity. Enthusiasm for progress was never greater: that was the age of steam, of railroads. Yet Comte's vision of progress was peculiarly his own. The end of theology. The end of metaphysics. Enter positive science. And now our conference. Shall we continue the story? The End of Science?

Comte was jubilant, if mistaken. Metaphysics, thank goodness, was at an end. The end of science is different. Our hosts, in their advance notices, declare their 'grave epistemological concern.' They did not invite us to debate the cheery question with which the great cosmologist, Stephen Hawking, progenitor of black holes, introduced his inaugural lecture a few years ago: 'Is the End in Sight for Theoretical Physics?' His title meant: have we done it, folks? Are we about to pull it off, tell the ultimate story about everything? On the contrary, it is clear that this conference is driven by fear of an extreme scepticism about science.

Just as Comte received mixed notices for his 'ends,' some people now hear 'The end of science?' with joy, others with distaste or even despair. (But as I shall say in a moment, all of us hear it with disbelief.) The joyful think we've had a belly full of science and should try something else. The despairing think we may be in for hard times now. The title question is also a pun. For you can hear it as asking not only, 'Is science coming to an end?', but also 'What is the goal of science?' Like so many puns, the two meanings can be made to play into each other. We might see our role today like this: we try to say what are the goals of science, or to define new aims. Once accepted as worthy, these become the end of science and stop science from coming to an end. I have no such aspirations. I do not believe in the pious picking of goals, and I do not believe that we choose our ways of life, our destiny or our styles of thinking. Our hosts ask whether we are 'at the end of science or at the beginning of new understandings that will legitimize science for the future.' Philosophers don't have a good record in the legitimation business. Hegel is often said (misleadingly) to have legitimated the power of the Prussian state; my heroes, from Socrates on, are those who revealed illegitimacies, and I don't want to join the other crowd, the gang of legitimators.

Nevertheless I am not a sceptic about science. I think that one of the grounds for scepticism is a mistaken self-conception--how off-duty scientists describe their own activity. Not surprisingly their unscientific friends have much the same conception of science. And it is false. It was forged for other purposes, in another era. So I'm in the 'legitimation' business to this extent: I think the tree of knowledge would be healthier if the excrescences were pruned. Hence I won't please those who hope that science is at an end, and want to throw the bastard out. They would be glad to hear the next chapter in Comte's story, telling how science too passes away. First theology, then metaphysics, now science. Their rejoicing would proclaim that a scientific age had been an essential part of our wars, our genocide, our threat to all life, our spoiling of the environment, our subjugation of most peoples and the oppression, in the name of a patriarchal conception of nature, of women. Science has been presented as the model of rationality, the triumph of reason, but we've seen where that got us. Let's eliminate science before it eliminates us. So goes what has been called 'the rage against reason.' To this I add something more pervasive. We are surrounded by extraordinary technological and medical products some of which we employ or hope to enjoy. The rage against reason is coupled with a deep suspicion of technology. Like Comte's vision of the human race, technology, it has been said, goes through three stages.[38] There is a honeymoon period when we delight in it, the 'mechanism as bride' syndrome. Then a second stage in which 'people are designed out.' Finally 'technology runs as anti-people.' Such technology arises from the imperative we have so strongly felt since the seventeenth century, to interfere in nature, to purify, stabilize and indeed create phenomena in the laboratory, and then to manufacture them by mass production. We should not distinguish pure science from such technology, but speak simply of 'technoscience.' And it is the 'technological imperative' that drives weapons research and hence the strategies of current planning for war.[39] Sheldon Glashow believes that knowledge and its uses are independent; I don't.

The rage against reason and the fear of technoscience do not exhaust the complaints. There has arisen a deep distrust of what might be called an ideology of science. That is the excrescence that I want to prune. But I disagree with the advance publicity for this conference. There I read that 'we have begun to think of science as . . . operating out of and under the influence of social ideologies and attitudes--Marxism and feminism, for example.' Marxism and feminism! Masculinism, maybe, but feminism? Some Marxists, not unduly faithful to Marx, are among those who are most critical of the presuppositions of science. The women's movement has provided some of the most cogent analyses of the entire practice of science since the European seventeenth century. Sandra Harding has been a pioneer as writer, organizer, editor and guide to a generation. She will

---

[38]I follow lecture 5 of Ursula Franklin's 1989 Massey Lectures for the Canadian Broadcasting Corporation, The Real World of Technology, to be aired this November, and later published.

[39]I take the term 'technoscience' from Bruno Latour's Science in Action, Cambridge, Mass.: Harvard University Press, 1987. I owe the phrase 'technological imperative' to Anatol Rappoport.

know all too well that only the most minute fraction of science practiced today has the slightest trace of feminist ideology. And precious little Marxism.

It is part of the power of 'science' that these counter-ideologies have so little effect. Science seems to have its own ideology, part of which is a fable about objectivity. Those who fear the hegemony of science say that it forces upon all people a conception that there is one truth, the scientific one. There is only one method, the scientific method. There is only one way to reason, only one way to reach valid conclusions. They say that what is presented as a value neutral assessment of the nature of truth, method and reason, is in fact an assertion of a particular set of values. It is an assertion that is both imperial and covert. It is made in such a way that we are not to notice how what is supposed to be the beginning of a set of questions into the nature of knowledge, has already determined in broad outline what the conclusions must be.

When I speak of the ideology of science, I perhaps misuse the very word 'ideology,' which, as Karl Mannheim had it, meant a system of belief that kept a practice in being. I am referring to the self-sustaining power of science, not to an ideology outside of it that keeps it going. Here I need two contrasts, one with the idea of form, and the other with the idea of interests. 'Form' is hard: I've never succeeded in clarifying it. Traditionally it is opposed to 'content.' It can be argued that some features of the form of scientific knowledge are, when looked at from sufficient distance, entirely contingent, while they are experienced by workers as a priori. We find the idea of an historical a priori in Foucault, who also for a time used the word 'epitome' more frequently.[40] Take even the idea that to find out about the world we need to find out how it works--a mechanical analogy leading on to the notion that an account of the world must be put in the form of 'this makes that happen.' Keller and others have urged that this constraint--or model for success--has historical roots in the scientific revolution, when Aristotle's four types of causes were reduced to one, 'efficient causation.'[41] What caused something was what 'made' it happen, and 'force' became the operative word in physics. This transformation is closely associated with the gendered metaphors of the time, of male dominance, of the master scientist making mistress nature his slave. You could well think of this as part of an ideology, as part of what sustains science. And it would be argued that this is entirely internal to science as it is now practiced. Even so, such a fundamental ideological underpinning of what we call science is at most a boon companion of what I wish to discuss.

As for interests: in Edinburgh and elsewhere there has arisen a school advocating what it calls 'the strong programme in the sociology of science.' Many of its early activists were students of Mary Hesse, and she has given one of its ablest characterizations.[42] It has urged that one can never explain a belief or

---

[40]The phrase is not Foucault's, but is aptly used by Georges Canguilhem to describe Foucault and then was picked up, with a wry smile, by Foucault himself.

[41]Evelyn Fox Keller, Reflections on Gender and Science, New Haven: Yale University Press, 1985.

[42]Mary Hesse, 'The Strong Thesis of Sociology of Knowledge,' Revolution and Reconstruction in the Philosophy of Science, Bloomington, Ind.: Indiana University Press, 1980, 29-62.

conclusion by its being true, or its being well grounded by the evidence. There must always be other causes analogous to those used to explain the maintenance of false beliefs and mistaken conclusions. In the early days of this movement it emphasized the role of interests outside of a science. An example is the enthusiasm of some members of the late-Victorian middle classes for eugenics, which arise from fear of the fecund poor, and which generated the statistical technology that we use today.[43] That is precisely what I am not talking about. I know that there is something of a continuum from such patently external interests to what I claim keeps science running on its own, to the very will to truth and objectivity itself. That does not matter to my present purpose. I am not talking about interests that serve any ends other than the maintenance of science itself. These may include the fundamental ethos of its male managers, and be embodied in a certain admissible form of scientific knowledge, but they do not include the interests of the Victorian middle class in controlling the fecund poor.

No End of Science

Before proceeding, I must pull up short. Whether you are an admirer of or skeptical about science, have no illusions. Reports of the incipient demise of science have been grossly exaggerated. Our invitation asks, 'Do we stand at the brink of "The End of Science"?' The answer is plain. No! What is striking about science is its near uniform acceptance, for better or ill. The proportion of the national treasure dedicated to fundamental research varies from time to time and from country to country. We disagree about how to distribute scarce resources like wealth and more importantly brains, but every person in any position of authority in the world wants some science, or at least pays lip service by saying so. Preferably lots of it, but please control costs and keep down the harmful side-effects. Four years ago, at a conference discussing creeping relativism about knowledge and values, Paul Feyerabend had to protest that aside from quite esoteric debates, relativism is not even on the cards, and that 'the basic phenomenon of "world culture" is the relentless expansion of Western views and technologies--monotony, not variety, is the basic theme of the age.'[44]

The one specific--aside from generic power itself--that people in other parts of the world wanted to get from Atlantic civilization is the 'science' inaugurated by the so-called scientific revolution that took place in a few European communities which, at that time, by world standards, were filthy, poor and weak. Science is increasingly a staple of the 'Pacific rim'; perhaps science has already moved its headquarters. Of course some other Western things are valued in many parts of the world, but none as much as science and the technologies that derive from it. I happened to be teaching in China April-June of this year, and so was pretty close to the pro-Democracy movement. I do not underestimate the importance of its ideals, for which at least one of my friends is in jail right now. But one thing was more important than democracy, and that was

---

[43]I refer to an analysis by Donald MacKenzie, Statistics in Britain 1865-1930: The Social Construction of Scientific Knowledge, Edinburgh: Edinburgh University Press, 1981.

[44]Paul Feyerabend, 'Cultural Pluralism or Brave New Monotony?' in Farewell to Reason, London and New York: Verso (1987), p. 273.

science. The high-prestige labs there continue apace. On the top floor of a cancer hospital, which by our standards gives poor care, there is as expert and innovative an advanced molecular oncology research laboratory as exists in the world. American technology in satellite cartography is adapted and improved to make what I am told are the most sophisticated and accurate maps in existence, doing for the vast empty expanses of Western China what has not yet been done for the American Midwest. Some of the most innovative work in mid-range temperature superconductivity continues in Peking. And so on. Democracy can be put on hold, but not science--and here we are speaking of science which, if not the 'purest,' is indefinitely far away from any application that will help improve the conditions of life in the third world. Westerners repeat: this is amazing research, but why are you doing this rather than something useful for your people? The answer, crudely, is we want to. To paraphrase Feyerabend, the relentless expansion of Western science and its self-conceptions is the basic theme of the age. There is no end in sight. But perhaps, as Gunther Stent has said in criticism of this remark, public enthusiasm and willingness to finance 'pure' science is running out, leaving us only with technology? There are straws in the wind: Thatcher's decimation of pure research in Britain, and the decision to restrict the vast potential of the German Democratic Republic to technological research. I would guess differently. No matter here. Such an end of pure science has little to do with revulsion against its ideology, but is instead founded on material greed and a theory about how best to satisfy it.

After this excursus I can return to my theme. The current anger at science, indicated by the title of our conference, is directed at an ideology of science that says there is one ultimate reality, one ultimate truth, one road to the truth (the scientific method), one sound mode of reasoning, one national way of speaking. There have been sceptics about such hegemony at least since Bacon and Galileo were perceived to vanquish the schoolmen. There has been angst in every later time, for example in the worries about science and religion that so troubled the soul of so many sensible Victorians. But at present the revulsion against science tries to undermine its pretensions to objective knowledge. To continue quoting from our hosts, 'science does not speak about extrahistorical, external, and universal law . . . there is no way of speaking of something real behind science that science merely reflects . . . A public aroused by visions of nuclear holocausts, genetic mutations and environmental collapse starts to question the products and process of science, as well as the reality that science is supposed to define and describe.' Now of course worry about the uses of science does not imply that science does not investigate reality; I would naively have thought the opposite, that our ability to harness the nucleus and then careen wildly to our own self-destruction shows that we do understand some aspects of reality, of the noncontextual, nonhistorical microworld. But there is not the slightest doubt that an antagonism to science fuelled by worries about its by-products, leads to a skeptical critique. 'What's so great about science anyway?' Hence the worry about its objectivity, and the alleged reality that it pursues. Now in listing complaints I said 'one' over and over again: one reality, one truth, one method, one rationality. In the initial triumph of 'science' in the seventeenth century, claims to oneness were radical denials. They were challenges by a minority of enthusiasts to an established order, and I in no way wish to put them down. Nor do I ever deny for a moment that people standing up against despotism and lies need those ideals of one truth, one reason, one reality. But incautious repetition of those ideals does science no good. So I intend to examine many things that can be captured in this seemingly innocent word 'unity.' In the end my target is the one that clearly most troubles our hosts, the wish for 'the reality that science is supposed to define and describe.'

'The unity of science' has, from time to time, been a slogan. It breaks into two parts: unity is a good thing, and science is or should be unified. The first is a judgment of value, the second an injunction. I shall begin with the value. Is unity a good thing? Not just the unity of science, but the unity of anything.

Is Unity a Good Thing?

Gerald Holton has written of what he calls themata. These are governing ideas about nature, humans or divinity that recur in the successive development of culture. Atomism, namely the idea of an elementary particle, is a prized example, from the speculations of Thales to the charm of quarks devised by Sheldon Glashow. Holton notes other themata: symmetry, homogeneity and conservation. His themata have at least three roles, as concept, as methodological precept, and as hypothesis. Themata are not peculiar to the sciences. Atomism is also a thesis about society. The western European conception of the person has been atomistic, an idea that now flourishes most in individualistic America. The eastern European--I include German--conception has been collectivist. In the one, the Hobbesian individual chooses the sovereign, the form of state control. In the other, the Herderian community determines the very nature of persona identity and conscience. It is like that with unity. Few pairs of themata have been more persistent than the one and the many. They infect, everything: God, people, knowledge, truth, beauty, empire. Your nickels and dimes have inscribed upon them, E pluribus unum.

God: when I was a child I was taught that one of the great triumphs of Western culture was monotheism, the one God of Israel transformed into Christianity and Islam. The battery of Greek gods was quaint. It is merely a symptom of our times that I no longer share that opinion. We no longer feel that unity is quite the perfection that the schoolmen made it out to be when they proved the existence of one and only one God. Some will interrupt me in fiercer tones: Unity of the Godhead has no other meaning than exclusion of all others, absolute hegemony, and patriarchal domination.

People: we have long learned that the healthy mind is a whole one, an integrated personality; all recent therapies teach that one must come in touch with one's feelings and be a whole person. I may well be an unhappy or torn person if my different selves battle with each other, but it is not obvious that the thing to do is to integrate them into one self. Why not have several collaborative selves? Why is that not a goal of therapy? So powerful has been the grip of unity on the Western mind that no one contemplates such 'cures'--no more than anyone preaches polytheism in the tents of a religious revival meeting.

Nations: empire has often been popular, for the imperialists, but unity as an intrinsically desirable goal for a nation may not be all that ancient or so enduring. It is characteristic of our times that that established piece of American ideology, 'the melting pot,' is no longer unquestioned. (The number of Swedish flags I can see from here--200?--suggests it was never a big value in Saint Peter.) E pluribus unum: we think that unity of the nation is such an eternal idea that this sounds like a Latin tag, doubtless from a republican senate in Rome. Not at all, it comes from the Reader's Digest--or rather, it was the motto of the Gentlemen's Magazine of London, 1692. The Magazine took excerpts from all the other magazines and printed them up as a single periodical for the busy Gentleman. Out of the many (periodicals), one. I shall later follow one of the great scientific unifiers, James Clerk Maxwell, in asking whether we should not hold the Book of Nature, as it has been called, to be like the Reader's Digest.

Unity as perfection impedes us everywhere. Everyone has asked, what is the meaning of life? And, failing an answer, in a moment of teenage, mid-life or senescent gloom, concluded that life has no meaning. We don't notice how the question makes a grand presupposition. Why should life have only one meaning? Why do we never revel in the many meanings of life? The unity value sometimes throttles us. I need hardly continue this theme into aesthetics and much else. Only a fool and a poet (wrote Nietzsche) praises diversity.

These frivolous remarks have had one purpose: to remind you that it is not inevitable that unity should have been thought of as a virtue. I now return to science, but notice one other thing: two distinguishable although interconnected ideas are at work in the theme of unity. Not even unity is at one with itself. The root word is unum, one, and for sure unity connotes singleness, oneness. I find it hard to think of oneness as a property of a thing. It is like existence; as Kant said, I do not add something to the golden dollars of the merchant in saying, 'and then they exist.' I do not add to the properties of an apple, after saying that it is crisp and tasty, that it is 'one.' but we can speak in context of unity being desirable: the unity of certain concerts that we have heard, novels that we have read. A speech, a political platform, may or may not have unity. This unity has something to do with the integration or harmony of the parts, a harmony that exists or does not exist after the item has already been individuated as one thing, one concert (starting at 8 and ending at 10:30, with an intermission).[45]

Let's call these two aspects of unity singleness and harmonious integration. The unity of the god of Israel and of Islam is singleness. The unity of the self is largely a matter of harmonious integration. The United States of America is a single nation, but the nation may not be united; the melting pot theory was part of an archaic theory about what makes for singularity? Or is the idea of oneness really something derived from an experience of harmony? Notice that some of the unities of science that I shall list have to do with singleness and some with harmony. Not even unity is united.

Eleven Unities of Science

The unity of science denotes at least eleven different families of theses. The first four are metaphysical. They correspond to Holton's themata as hypothesis and form an 'over-arching meta-scientific hypothesis.' The next four are epistemological; they are about the aim of knowledge, what to find out, and what knowledge should look like. They are Holton's methodological precepts. The next two are logical in nature, and the last one is historical, asserting that the methodological precepts have been delivering the goods: unity is in good shape, thank you very much.

---

[45]I owe the distinction to a talk on the soul by Elizabeth Anscombe, in Toronto, 28th September 1989.

1.   There is only one world, one reality, one truth.   That is the metaphysical thesis or bundle of theses at the top of the list.   Put baldly, it is a little hard to know what it could mean--we don't go around counting worlds.

2.   For some scientists it has meant that all kinds of phenomena must be related to each other.   Faraday, for example, believed that the world could not be such that light and magnetism did not somehow affect each other, and spent two decades trying to find out how.   In the end he showed that magnetizing could change the polarizing properties of some substances.   Glashow and his colleagues were convinced that the strong, weak and electromagnetic forces must have some foci of interaction.   This is a thesis of interconnectedness.   To my mind it is the clearest of the metaphysical theses.

3.   Philosophers like to refine the idea.   The world, as Wittgenstein put it, may be made up of facts, but it is not a ragbag of facts.   There is a unique fundamental structure to the truths about the world, with central truths that imply peripheral ones.   Metaphysicians would say that this is a structure of causes, necessary of probabilistic; positivists, who don't like causes, would say it is a structure of logical relations between laws.   However qualified, this is a structural thesis.

4.   In ordinary life we classify in ever so many different ways, but there is an underlying belief that there is one fundamental ultimate right system of classifying everything: nature breaks into what have been called 'natural kinds.'   This is a taxonomic thesis.

5.   Now we pass to epistemology which will as usual recapitulate the metaphysics.   We can never know the whole truth about everything, but all our pieces of knowledge are fragments of the truth.   Science aims at knowing the truth about (aspects of) the one world.   This is the core epistemological thesis or bundle of theses.

6.   The interconnectedness thesis suggests the injunction followed by Faraday: elicit, in experiment or theory, connections between types of phenomena.   He did so because he thought God would not have made a world in which phenomena bore no relation to each other.   The legitimate desire to find Grand Unified Theories is driven by this precept, which often runs into or is confused with the next, which is much less clear.

7.   Attempt not only to find out the truth, but also the structure of truths.   Philosophers, obsessed by logic and language, like this idea.   Many enjoin us to reduce the laws of one body of knowledge to the laws of another, in particular, economics to sociology, sociology to psychology, psychology to biology, biology to chemistry, geology to chemistry and geophysics, chemistry and geophysics to physics, and then unite physics and cosmology in a Grand Unified Theory.   This is a reductionist thesis.

8.   Logical positivists, who were much preoccupied with the language of science, believed that the reductionist thesis required that all of science could be expressed in one language, for how else could the theory of one discipline be reduced to that of another?   This is a linguistic thesis, loosely related to the taxonomic thesis in that it implies that all our systems of classification should mesh.

9.   There is only one standard of reason by which scientific hypotheses can be judged.   This is a thesis about rationality, and it is a logical thesis.

10. There is one best way to find out about the world and how it works, the scientific method. The same method is to be used in all the sciences, natural, social and human. This is a methodological thesis.

11. Science has been wonderfully successful not only in finding out new facts about the world and in creating and controlling new phenomena, but also in its unifications, bringing many facts under the wing of one intellectual structure. Light, optics, radiant heat, electricity, magnetism, were brought under electromagnetism. Molecular biology has proved profound connections between organic chemistry and genetics. Just as high energy physics was about to drown in a plethora of its own particles, gauge theory ordered them in unified principles. this is an historical thesis about the success of projects of unification.

Reality and Objectivity

I shall say nothing of Reality, the World, or the thesis that Being is One. As I said overly sardonically, we don't go around counting worlds. I should nevertheless mention one 'Reality-linked' reason for philosophical enthusiasm for the unity of science, current in the 1920s. That enthusiasm founded the Unity of Science Association, and projected an Encyclopedia of Unified Science. Its founders were closely associated with the Vienna Circle, although the last volume in the Encyclopedia was none other than Kuhn's unpositivist The Structure of Scientific Revolutions. Now if we look at those men in the heyday of logical positivism, we shall find that many were thorough-going phenomenalists. Sense-data were primary reality. For Carnap and his colleagues, there was only one royal road leading beyond the veil of appearances, and that was science. But if science were not united, we would not be led to one world. The thought that there could be more than one world was repugnant to the bluff commonsense that underlay positivist phenomenalism. So there had to be a unified science.[46]

There are two confirmations of this connection between phenomenalism and unity. The great successor to Carnap's Aufbau (his attempt to 'construct' the world) was Nelson Goodman's The Structure of Appearance. As the great American pragmatist matured, he did not abandon phenomenalism, but instead came to think of it as just one possible version. Not 'version of the world,' for Goodman describes the world as 'well lost.' Realizing that a phenomenalist requires the unity of science to be confident of one real world, Goodman has abandoned unities root and branch. The lopped branches show everywhere in his philosophy; at the root he has cut out the very first unity thesis, the world itself.

Moving in the other direction in time from Carnap, Kant's transcendental idealism is full of demands for unity, none more stringent than the transcendental unity of apperception. Kant believed that there was no possibility of objective knowledge without a number of such unities. Philosophers have not been the sole source of the quest for unity. But they certainly have been there to support the search of the religious scientist, trying to show that all God's manifestations are interconnected.

Turning from pure philosophy to science, the elevenfold unities do not exhaust scientific ideology, but they do bite off chunks of its metaphysics, the theory of knowledge, and of logic, method and history. That is quite enough.

---

[46] I owe this observation to Judith Baker.

Apparently I have omitted the claim to objectivity, and to discovering interpersonal value-neutral truth. In fact the metaphysical theses say that there just is one body of truth, one reality, regardless of what people think. The supposition of objectivity pervades the epistemological theses. All fragments of knowledge are to be interconnected in one objective deductive structure. The logical thesis says that there are objective good reasons, evidence. The methodological thesis says that there is just one objective way to engage in research. Even the historical thesis makes a claim to objectivity: there is a body of something like the organization of the truth that we have found out.

The Book of Nature

Objectivity does not imply unity. James Clerk Maxwell never doubted objectivity, but he did think that unity could be called in question:

> Perhaps the book, as it has been called, of nature is regularly paged; if so, no doubt the introductory parts will explain those that follow, and the methods taught in the first chapters will be taken for granted and used as illustration in the more advanced parts of the course; but if it is not a book at all, but a magazine, nothing is more foolish than to suppose that one part can throw light on another.[47]

The metaphor of the Book of Nature is an old one, much used by Galileo when he spoke of the Author of Nature writing in the Language of Mathematics. I like it best as a version of Leibniz's vision of knowledge and the world. It makes excellent sense of the eight metaphysical and epistemological theses of unity. Leibniz thought of the world brought into being by God, who contemplates the ideas of all logically possible worlds to determine which is best. It is a very small jump to making Leibniz's God not only supremely reasonable but also supremely wordy, a God who contemplates all possible world-descriptions and then creates the most perfect world answering to a possible description. The Book of Nature provides just that description. It is clear that Leibniz thinks it is essential to such descriptions that they have structural, reductionist and taxonomic unity. He sees science as imitating God's work, and thus as at least aiming at a best language.

Admirers of metaphysical unity should compare themselves to admirers of metaphysical simplicity. Theories and laws are often thought to be better if they are simple. I well understand a preference founded upon aesthetics of ease of computation, but suppose it is also urged that simple theories are more likely to be true. Why? Leibniz had a ready answer, that God preferred the simplest theory with the most diverse consequences; it was elegantly economical. I know of no other reason for thinking simplicity a guide to the truth. I suspect that many admirers of unity have, au fond, a thoroughly theological motivation, even though they dare not mention God. I wish they would! It would get things out in the open. And that is one of many reasons for being grateful to Mary Hesse's Gifford lectures with their comparisons between science, myth and theology. Comte's

---

[47]L. Campbell and J. Garnett, The Life of James Clerk Maxwell, London: MacMillan (1892), p. 243.

bland confidence that we are out of the theological era is not fully justified so long as people maintain the myths of the unity of science.

Leibniz's charming story is far from antiquated. Only in terms like those of Leibniz can I understand the most widespread picture of contemporary cosmology. In the 'big bang' account of the universe, it is thought that there are certain fundamental laws of nature, in which occur laws with certain parameters, fundamental constants such as the velocity of light or the gravitational constant. In one line of discussion that has acquired some notoriety of late--the cosmological anthropic principles--it is noted that most choices of values for those constants would lead to very uninteresting consequences. The universe would quickly collapse or explode. Indeed if the constants aren't just right, heavy metals would not form, galaxies would not cohere . . . and so on. So a question is posed, how come the constants are just right? I cannot understand this question outside of a picture of first the laws of nature being written down in God's sketch book, with free parameters, and then the blanks being filled in with particular constants of nature. More generally, I find it very difficult to grasp the force of the theses of the unities of the sciences, without having in the back of my mind that wondrous image, the Book of Nature. And Maxwell's irony undoes it all in a paragraph. If you must picture the world in a wordy way, why not imagine that God edited a magazine?

Trends Towards Unification and Towards Disunity

I don't despise unity, especially in the sense of harmonious integration. Some of our noblest intellectual achievements have been unifying ones. I happen to like experimental work more than most recent philosophers, and so one of my heroes is Michael Faraday with his magneto-optical effect, the first phenomenon that made plausible that there could be a unified theory of light and of electromagnetism. The unifying power of the work for which Weinberg, Salem and Glashow shared a prize is remarkable for any era; one can say, without hyperbole, that it brought high-energy physics back together again. More familiar to most of you, because part of every high school science curriculum, has been the achievement of molecular biology in creating bridges between organic chemistry on the one hand, and genetics on the other. Unification <u>works</u>!

But be cautious. Sheldon Glashow remarks to me that only in the most vicarious way can quantum chromodynamics and electroweak theory be said to be unified. Biology and chemistry are now unified? What has Gunther Stent been doing in the past couple of years but presiding over a vast restructuring of the Berkeley biology and life sciences organization, breaking up groups and remoulding others in the light of the changes of the past twenty odd years? The upshot is not a unified department of biology. He is head of the department of molecular biology, which has six divisions which he himself did not want to bring under one super-department. Across campus is now a disjoint super-department of life sciences. Microbiology and macrobiology have been institutionalized further apart than ever!

Unity is no simple matter. Too much talk about the unification of the sciences is unreflective gossip. Hence I warmly commend the work of a number of philosophers who have attempted to say in what ways we have of late achieved various unifications and reductions. A quite easy and very informative read is Alexander Rosenberg's <u>The Structure of Biological Science</u>. He makes rather plain what has been reduced, and in what sense it has been reduced--and how much has not been reduced, and may in several senses be called irreducible. The questions about nuclear physics are of course more difficult to follow, and perhaps more difficult to analyze; indeed it requires tact and tenacity to state clearly in what sense Maxwell unified electricity and magnetism.[48] I strongly encourage careful attempts to study unification and reduction. The best among them imply neither that there is one kind of unity, nor that <u>all</u> science has even <u>one</u> of my eleven kinds of unity.

These cautious remarks do not cast aspersions on the historical thesis (11), that various kinds of unification have in fact been achieved. Is that not excellent inductive ground for most of the preceding ten theses? The reason that we have been able to unify is that there is one real structure of truths that can be discovered by our logic and our method. But there is a worry, implicit in one of the wisest of essays favouring unity, 'Unity of Science as a working Hypothesis,' published over thirty years ago by Paul Oppenheim and Hilary Putnam.[49] They assert the historical thesis of ongoing unification as 'a trend within scientific inquiry,' to which they quickly add, 'notwithstanding the simultaneous existence (and, of course, legitimacy) of other, even <u>incompatible</u> trends.' They don't explain, but we know what they mean. We hear endless complaints about overspecialization of the sciences, about myriads of papers in journals that nobody reads. In a nice book, <u>Probabilistic Metaphysics</u>, Patrick Suppes has well argued the case for what he calls 'the plurality of science.' He starts just with trying to read a paper in a journal to which his daughter, a student neurophysiologist, subscribed. He observes, in a droll way, how the language is just inaccessible ('if postsynaptic adrenergic neurons in neonatal rats were chemically destroyed with 6-hydroxyudopamine . . . the normal development of presynaptic ChAc activity was prevented.'[50]) The point is not that you can't learn to read this, nor that there is any problem in mastering any one body of experimental technique. The point is that in a quite straightforward sense there is no common language of science, and that as a matter of practicability, there could not be. In 1800 it was true that an alert person, a Thomas Young, perhaps, could savour every article in the <u>Philosophical Transactions of the Royal Society of London</u>. That is no longer humanly possible, and that is a fact about, among other things, language. And is this a bad thing? Suppes applauds

> the divergence of language in science and [he] find[s] it no grounds for skepticism or pessimism about the continued growth of science. The irreducible pluralism of languages of science is as desirable a feature as is the irreducible plurality of political views in a democracy.

---

[48]See Margaret Morrison, 'A Study in Theory Unification: The Case of Maxwell's Electromagnetic Theory,' <u>Studies in the History and Philosophy of Science</u>, forthcoming.

[49]Paul Oppenheim and Hilary Putnam, 'Unity of Science as a Working Hypothesis,' in H. Feigl et al., eds., <u>Concepts, Theories and the Mind-Body Problem</u>, Minnesota Studies in the Philosophy of Science II, Minneapolis: University of Minnesota Press (1958), 3-36 on p. 6.

[50]Patrick Suppes, <u>Probabilistic Metaphysics</u>, Oxford: Blackwell (1984), p. 121.

And what about the reduction of theories? Even 'quantum chemistry, in spite of its proximity to quantum mechanics, is and will remain an essential autonomous discipline.'

It is hopeless to try to solve the problems of quantum chemistry by applying the fundamental laws of quantum mechanics . . . The combinatorial explosion is so drastic and so overwhelming that theoretical arguments can be given that, not only now but in the future, it will be impossible to reduce the problems of quantum chemistry to problems of ordinary quantum mechanics.

The comparison is to a chess-playing programme. In principle you can't programme a computer to 'foresee every possible move' in a game. There are about 10^{120} possible moves in a game, perhaps 10^{40} times as many atoms as there are in the universe.

If superstring theory or some other turns out to be the most powerful of grand unified theories, would it affect the centrifugal tendencies? Should it? In the inaugural lecture of Stephen Hawking that I mentioned, he is cheerful to say that even if the end is in sight for theoretical physics, all the rest of physics remains to be done, forever.

Method and Reason

We may feel more comfortable turning to the claim that there is just one reason, and one scientific method. Yet these are somewhat hollow claims. Take scientific method first. Let us turn to an ingenious piece of research by members of a team directed by Lap-Chee Tsui. After seven years of unceasing toil, they have identified the genetic material that carries most cystic fibrosis. If you ask, what (scientific) methods did they use, the answer is impressive. I shall quote not from a research paper but just from common journalism, the New York Times (September 12, 1989): genetic linkage analysis. Pulse field electrophoresis. Chromosome jumping, a method that they pioneered. Saturation mapping. Recombinant work. Zoo blotting (you compare some isolated DNA with that of various animals to see if they have it too, if so, you think it might be genetically important). Then endless work with a copy DNA library. Finally polymerase chain reaction. It would be silly to say that they used Zoo blotting, polymerase chain reaction, and the scientific method. Where then is this splendid specific of science, the scientific method?

We might try the following. Husserl spoke of the Galilean style of reasoning, of making abstract models of the universe, whose phenomenological consequences we explore. The idea is expressed more strongly by Weinberg, who recorded how 'the physicists, at least, give a higher degree of reality than they accord the ordinary world of sensations.'[51] A remarkable style, since 'the universe does not seem to have been prepared with human beings in mind.' and this comment has been used by Noam Chomsky as part of a legitimation of his approach to linguistics, which he says uses the same style, and he writes that, 'we

---

[51]Steven Weinberg, 'The Forces of Nature,' Bulletin of the American Academy of Arts and Sciences 29 (1976) p. 28.

have no present alternative to pursuing the "Galilean style" in the natural sciences at least.'[52] Might this not be the very scientific method, which some would say originated in the time of Galileo? Might we not say that the Tsui and his colleagues had used the Galilean style, first postulating that there was a certain reality there, a reality of a certain form, a genetic carrier of a certain type for an apparently hereditary disease? What we wanted for our thesis of methodological unity was a specification of the method, and that is precisely the Galilean style!

A moment's reflection won't let this stand. It does not begin to capture what Tsui and his group were doing To turn to a different level of generality, just when Weinberg and Chomsky wrote in praise of the Galilean style, a distinguished historian of science, A. C. Crombie, was writing about styles of scientific reasoning in the European tradition, among which he enumerated six.[53] In brief, (a) postulation in the axiomatic mathematical sciences, (b) experimental exploration and measurement of complex detectable relations, (c) hypothetical modelling, (d) ordering of variety by comparison and taxonomy, (e) statistical analysis of populations, and (f) historical derivation of genetic development.

Are all our unities expanding into plurality? There are still quite a few of my eleven left to go, but I shall call a halt. Briefly: Unity (4) was taxonomic. I do think there is a good sense to the idea of 'natural kinds,' but in explaining it I reject my call for a single ultimate taxonomy that obscenely, as philosophers from Plato on have been putting it, 'cuts nature at the joints.'[54] Closely related is the idea that there is one fundamental structure of causation, deterministic or probabilistic, that runs the world. The idea that causation is like that is truly a residue of the ages that Comte optimistically hoped had passed away; today I would say that it rests upon an improper analysis of 'cause.' I don't say it always did, for our conception of causation has been travelling along quite a few roads since the time of Leibniz, say. The one type of unity that seems to be both clearest and most important is that of connectedness. It always makes sense to ask if this is connected to that. Sir George Darwin said one should from time to time do bizarre experiments, such as blowing the trumpet to one's tulips every morning to see if they prosper better. It is not to be expected there would be such a connection, but it would be glorious if there were.[55] Connection, interaction and collaboration of aspects of nature remains an ultimate object of human curiosity

---

[52]Noam Chomsky, Rules and Representations, New York: Columbia University Press (1980), p. 9.

[53]I encountered this at a conference in 1978 and then in a book that Crombie had then almost completed. I adopted this talk of 'styles' in my 'Language, Truth and Reason,' Rationality and Relativism, M. Hollis and S. Lukes, eds., Oxford: Blackwell, 1982, 48-66. Crombie's book has still not appeared. For the list of styles see A. C. Crombie, 'Philosophical Presuppositions and Shifting Interpretations of Galileo' in Theory Change, Ancient Axiomatics and Galileo's Methodology, J. Hintikka et al., eds., Dordrecht: Reidel (1981), p. 284.

[54]Ian Hacking, 'A Tradition of Natural Kinds,' forthcoming in Philosophical Studies.

[55]Reported in J. E. Littlewood, A Mathematician's Miscellany.

and wonder, whatever we think of those airy notions of reduction, causation, justification, logic or methodology and all the other lazy myths of unity.

Unifiers

Is science then one kind of thing at all? There is no set of features peculiar to all the sciences, and possessed only by sciences. There is no set of necessary and sufficient conditions for being a science. There are a lot of family resemblances between sciences. Importantly there are quite different kinds of 'unifiers'--that is an interesting theme that I shall not develop. The first of the unifiers of modern science, mathematics. During the nineteenth century the differential and integral calculus were the trademarks of much science. Analysis and its descendants governed science for the first half of the present century, be it population statistics or nuclear technology. But there are many mathematics that unify. Quantum electrodynamics became a unified theory, perhaps, only with the advent of the technique called renormalization. Electroweak theory became the paragon of unification precisely when it was cast in the form of lagrangians, an eighteenth century structure determining the success of an achievement of the twentieth century. Once we start looking closer at these real-life examples, the kindly common denominator 'mathematics' ceases to denote one thing. Wittgenstein's phrase, 'the motley that it does such a good job of making much science look as if it were one unified activity: if we can apply mathematics to it, it must be one thing!'

There are many more unifiers, that is, tools, practices and bodies of knowledge that span sciences. In our day fast computation is the thing. It works at both sides of the experimental/theory divide. On the one side we can articulate theories that a decade ago would have had few intelligible or phenomenological consequences. On the other, data that no multitudes of human beings could ever have processed are transformed in short order into meaningful results. And it is not just a matter of calculating results, but of image enhancement, and of the design of apparatus, for example, in the way in which an old fashioned reflecting telescope is surpassed by many bits of the telescope, kept in exactly the right place, free of surface distortion of the sort that plagues enormous pieces of glass buffeted by winds and stressed by temperature change. Indeed to a certain extent computation replaces much experimentation by at least eliminating many possible scenarios: recombinant genetics and the acoustic architecture of the theatre will increasingly be done at the console, at most stages until the final prototypes are called for.

Less readily noticed is the way in which scientific instruments are among the unifiers of disparate bits of science. Throughout most of our century regimes and practices of experimentation and instrumentation have been more powerful a source of unity in science than grand unified theories. Instruments are speedily transferred from one discipline to another, not according to theoretical principles but in order to interface with phenomena. The scanning tunnelling electron microscope is a relatively new device that uses an effect called quantum tunnelling to produce sharp images. At first, like the transmission electron microscope, it was thought suitable only for metallurgy, but it expands into cell biology in ways not all that well thought out, and sometimes by accident. Very few of the consumers of this device understand how it works. For that they rely on facilitators. An extreme literally 'instrumentalist' thesis suggests itself: it is not high level theory that has stopped the innumerable branches of science from flying off in all directions, but the pervasiveness of a widely shared family of experimental practices and instrumentation.

Experimental Stability

Very well. Suppose that none of the eleven unities of the sciences are as entrenched as some have thought. Nevertheless, is not unity an ideal, a Nobel conception? Disunity may be a fact of life but a distasteful one. Does it have any positive virtues? Yes. I shall now argue for another kind of disunity taken from the notebooks of physics.

In his address to this conference Sheldon Glashow defends traditional fragments of science against know-nothings. Newton may not be the last word, but at least, he says, Newtonian mechanics is valid in its domain, as was classical quantum mechanics and numerous other instances. This is the germ of Werner Heisenberg's philosophy of nature. Even in 1926--before his new matrix approach to quantum mechanics had been mastered--he was saying that physical theories are essentially 'closed.'[56] As he explained the idea later, 'with that degree of accuracy with which the phenomena can be described by the Newtonian concepts, the Newtonian laws are also valid.'[57] The trouble with the Newtonian vision, and that of all classical physics, was that it pretended to be a universal system. That had been the root of Goethe's bitter antagonism to Newton.[58] And I would add that although of course "Newton" has become as much the name of an idea as of a man, Heisenberg's history-sketch was surely right. The Old Aristotelian notion of 'natural philosophy' did not assume or aim at the one ultimate theory of everything. That came only in the seventeenth century, whose culmination is displayed in the very title of Newton's masterpiece, which by implication identified mathematical principles and natural philosophy.

The idea of a closed theory with its domain at once suggests disunity: different domains governed by different theories. The authority of Heisenberg and other writers is no argument. They wrote from an exciting time of upheaval in physics, and were also much governed by positivistic principles. But I believe that their position helps us understand why (despite what T. S. Kuhn has taught us about scientific revolutions) so much of laboratory science is stable. It is not because there is a domain of experiment, given by nature itself, to which certain

---

[56]The term for 'closed theory' was 'Abgeschlossene Theorie.' 'Mehrkorperproblemen und Resonanz in der Quantummechanik,' Zeitschrift für Physik 1 (1926): 411-426. 1 (1926): 411-426.

[57]'Der Begriff "Abgeschlossene Theorie" in der modernen Wissenschaft,' Dialectica 2 (1948), 331-6. Physics and Philosophy: The Revolution in Modern Science New York: Harper and Row, 1958.

[58]Wandlungen in den Grundlagen der Naturwissenschaften, Leipzig: S. Hirzel (1945).

theories are true. It is because there is a mutual maturing of types of apparatus, phenomena and theory.[59]

To grasp this we must stop thinking of laboratory work as probing a reality that is inert, given. Most phenomena investigated in the laboratory do not exist, at least not in a pure state, prior to the intervention of the experimentalist. Experiment can uncover or reveal phenomena, but it can also literally bring them into being. (The handiest but by no means untypical example is the laser: nothing lased before we created this phenomenon in the laboratory). Our theories must answer to phenomena purified or created in the laboratory. But who answers whom? We have all learned that we revise theory to fit experimental results. Likewise we revise our interpretation or even our reading of observations in order to fit theory. There is a mutual adjustment. Excellent, but such talk of theory and observation leaves out the material world of the phenomena that we create and the apparatus with which we do it. There are not two malleable resources (theory, interpreted data) but at least five.

At the upper and more theoretical end, there are the rational, often mathematical, superstructures within which individual theories are embedded. Even these are revisable, but not easily. At the lower end we should long ago have added the physical devices that we design and use and modify; we should increasingly add the computational routes by which we reduce data. In between high theory and summaries of data, there is our experimental apparatus and our account of how it works, what it does. There is also the battery of procedures by which we make our records intelligible, perspicuous. We modify all elements in order to bring them into some kind of consilience. When we have done so: we have not read the truth of the world. There were not some pre-existing phenomena that experiment reported. It made them. There was not some previously organized correspondence between theory and reality that was confirmed. The theory is true to the phenomena that were elicited by instrumentation in order to get a good mesh with theory. This is a process of modifying apparatus both materially (we fix it) and intellectually (we redescribe how it works). It furnishes the glue that keeps our intellectual and material world together. It is what stabilizes science.

Such stability requires the disunity of Heisenbergian closed systems. Stability results from a sort of self-authentification resulting from the mutual adjustment of theory, apparatus, data and much more. We get stability across radical change partly because a great many lesser scientific revolutions do not result in discarding a body of knowledge but in supplementing it with new kinds of instruments, creating a new category of data for which radically new theory is demanded. We do not discard geometrical optics or Galilean mechanics; they fit the data furnished by instruments in their domain. But other theory is true to data provided by more subtle instrumentation. Here we see a new use for Kuhn's idea of incommensurability. Geometrical optics is incommensurable with wave theory because it is true to different phenomena. In this conception, the stability of the laboratory sciences depends upon their disunity.

Not that we need to be recherche to get the right sense of disunity. Eddington used to begin an enthralling lecture by saying that there were two

---

[59]This is an adaptation of ideas of Andrew Pickering, which I elaborate in a paper, 'On the Stability of the Laboratory Sciences,' forthcoming in A. Pickering (ed.), Science as Practice and Culture.

tables in front of him, the one solid, wooden, the other atomic, with vast spaces between the particles. Philosophers ridiculed him: there is only one table! Yes, but he was on the right track. There are different and incommensurable data domains, susceptible of study by and interference from different instruments.

Interference or Collaboration?

I have been talking of the creation of phenomena and emphasizing the way in which all the elements of laboratory science are moulded: the phenomena are brought into being, our theories about the instruments are altered, new instruments are made, rational superstructures are replaced--and so on. The philosophy of experimental science is vastly more lively than that leaden dead-weight, a philosophy of pure theory that has long obsessed philosophers of science.

Theory-oriented philosophy gave us a picture of nature as not only some unified totality but also as passive and inert. It made us think that we discover her properties, we reveal her secrets. It is some progress to speak, as I have done, of the experimenter creating phenomena. But still my picture has been one of interfering with nature in order to produce new effects. That is Francis Bacon's precept for the new science: we should twist the lion's tail in order to extract her secrets. As writers from Carolyn Merchant on have been showing by ample quotation, the Baconian metaphor is that of the master scientist commanding mistress nature.[60] Perhaps the scientific image of the future will be one of the experimenter collaborating with nature rather than controlling it.

This tendency will assuredly be augmented by the fact that the most imaginative sciences of the present decade are biological and astrophysical, one a vast bundle of life sciences and the other not a laboratory science at all. The most fashionable enquiry today may be chaos theory, topic of next year's Nobel Conference. We assess it differently. Gunther Stent suspects that it might prove to be the end of science, since it abandons attempts at explanation, or so he supposes. I am skeptical because chaos theory has had too much type, much like catastrophe theory a decade ago. But at any rate it tells a story of disorganized events of a sudden manifesting structure without being 'forced' to do so. And experiments in chaos theory are different in kind from ones that are more familiar to us. Physics has long lived by the following picture of an experiment: there is a target, some apparatus used to interfere with the target, and a detector used to determine what is the effect of the interference. Those are once again the words of James Clerk Maxwell. If you think I exaggerate the military overtones of targets, recall Rutherford's imagery on first splitting the atom; he compared his alpha particles to shells from a 19 inch gun, at that time the noblest achievement of the Royal Navy. The more biology liberates itself from a desire to emulate physics, the more the physicists' conception of their work may in turn be modified.

That would involve a change in our relation to and interaction with nature, but it could also change our very idea of natural processes. The Baconian image of the man-scientist interfering with the woman-nature has been projected on to nature itself. The sciences have been preoccupied by a model of central causal structures that dominate everything that happens. One can speculate that if an

---

[60]Carolyn Merchant, The Death of Nature, San Francisco: Harper and Row, 1980.

image of experimenter as 'biological' collaborator were to take hold, we might not only do science differently, but open our imaginations to new possibilities about how things happen in nature. We could come to think of the autonomous and independent activities of nature in a different light, not as run by triggering mechanisms and the like but by cooperation.

I have been gentle, in this essay, to only one kind of unity, namely the idea of connectedness. Different aspects of nature, that seem at first so unrelated, can be shown to interact. This type of unity gracefully shifts from a picture of one phenomenon governing another, to a vision of various phenomena collaborating in the production of a whole. I hope that the old ideology of unity will fade away, along with its implied meta-ideology that there is really only one kind of unity. What will be left is the idea of connectedness. Hence we will still admire both Faraday and the inventors of quantum electrodynamics.

The End of An Ideology?

Science has from time to time served as a model for all culture, most notably in the Enlightenment, but also during the nineteenth century when the confrontation between religion and science so troubled reflective Europeans. At present it does not seem a model, but that is because science itself is so ill understood. The humanities have clung to the Enlightenment image of science as a grand unifying intellectual adventure, one that strives to find the ultimate theory of everything. Our civilization now values accommodation, variety, choice. It denies foundations but yearns for a stability that ensures coexistence of a multitude of interests. It wants toleration and respect, not unified hegemony. As the human sciences have become more and more diversified, as writing and composing and dancing and designing have become more varied, science had been cast as a stereotyped monolith. But science has become as multifloriate as the humanities. It has become a domain in which there can be stability without foundations, sharing without commensurability. It is a domain that favours realism about the material world, with a maximum of variety but a minimum of subjectivism. It has become a domain in which there can be coherent action within a thoroughly disunified world picture. As we diminish the importance of the ideal of unity, we simultaneously treat theory as a partner with experiment. So we may be led to a modest experimental humanism, one which will be rooted, perhaps, more in the life sciences, than in physics, that old bastion of theory and of unity. That is the end of neither science nor theoretical physics. It would be a continuation of Comte's 'progress,' paring away more layers of a Leibnizian theological metaphysics. It might lead us to the end of an ideology of science that we are outgrowing. But that would not be not the end of science. Perhaps it is just what science needs to sustain itself in the face of the rage against reason.

QUESTION

ELVEE: Here is a question from our audience. If experiments create phenomena, what determines which succeed and which fail?

HACKING: Somehow I don't want to say 'there are all those laws written down and they determine that our apparatus would work or won't work.' I want to adopt a much more phenomenological approach. The only thing we can talk about are the phenomena to which the models and theories which we have are true. To what are laws and theories true? They're not true of the world. They don't reflect the world, but they are true to the phenomena that are measured, and that is the world in which our theories live, and where they become stable, by what I call that glue of the way in which we move around all sorts of things. So, it's not that there are no constraints, it's just that there's the constant need we've got to say those constraints must be laws which were there long before we got to making the phenomena.

Need a Constructed Reality Be Non-Objective?

Reflections on Science and Society

MARY HESSE

Previous speakers have given diverse, challenging and controversial interpretations of the "end of science." Of the many meanings of the word "end," some still need drawing out in more detail, and three of the issues arising from these will form an agenda for my contribution.

1. I shall start by looking at the multiple ambiguities of the title of the Conference. I don't know whether science as we know it, in its sense of "natural science," will shortly come to an end in history. Undoubtedly it will eventually, in a new Dark Age, or a holocaust, or in sheer social disinterest, but it is for prophets, historians and journalists to discuss whether this is imminent. What concerns me more immediately is whether we have come to the end of a conception, image or myth about science, and what is or should be replacing this. Closely connected is the question, what is or should be the goal of science, a question that involves the notion of objectivity that was raised in the introductory material for the Conference.

2. Were the positivists and instrumentalists right after all in making a distinction, perhaps fuzzy but nevertheless a distinction, between the experimental and the theoretical ends (goals) of science, and may the answer to questions about objectivity be different with respect to experimental and theoretical knowledge? This raises further problems about the theoretical language of science, because theories come typically in models and metaphors, not in direct observational descriptions.

3. How does science in the Anglo-American sense (Naturwissenschaft) relate to Scientia, Wissenschaft, or knowledge in general, in the European tradition? Is natural science value-free, or determined by a limited and partial set of values? How are we to judge other claims to knowledge (social science, history, theology) that have different value involvements and different kinds of constraints on their acceptability? Can values themselves be objective?

A big agenda, but one that would have to be addressed to give any sort of adequate response to the question raised in the Conference title. This was also the agenda for a series of Gifford Lectures given by Michael Arbib and myself a few years ago, which were published under the title of The Construction of Reality. There we tried to trace the relations and the continuities between knowledge of nature, of persons, of society, and ultimately of God, in response to the terms of Lord Gifford's foundation of his Lectures in 1885. He asked that the science of Infinite Being should be treated "as a strictly natural science . . . considered just as astronomy or chemistry is." In the 1980s, of course, we had to notice that his terminology implies the sort of imperialist view of natural science that the planners of the Conference rightly assume has come to an "end." So what was needed was a critique of this traditional account, to make room for nothing less than a new conception of knowledge that could better accommodate knowledge of nature and of God, and also of those important intermediate entities: persons and their societies, that have tended to be overlooked during the centuries-long controversies between so-called "science and religion."

Michael Arbib is a computer scientist, whose own description of his religious position is that he doesn't know whether he is an atheist or an agnostic. I am a philosopher of science who is a member of the Church of England. But in spite of our different starting points we found ourselves in immediate agreement about the importance of the new sciences of Artificial Intelligence and human cognition for understanding the nature of knowledge. The science of the brain is in its infancy, but attempts in AI to develop computer simulations of human and animal cognition do already give hints about a new way of thinking about knowledge in general. Let us start with the way we learn to perceive and manipulate things in the external world.

Arbib's basic concept for his models of brain-function is the schema: a unit that may refer to an object in the world, or more generally to an action, for example the grasping of a cup of coffee. Note that this is already a much more complex notion than the usual starting point of a theory of knowledge. It does not presuppose an external object, the cup, to which some concept or idea in the head refers in a one-to-one atomistic fashion. It does not distinguish an independent pre-given goal "picking up the cup" drawn from some other portion of the head labelled "purposes." It does not talk about the muscular movements and manipulations of external objects as if these were just tacked on to objects, perceptions and goals, and have to be triggered at some subsequent second in order to realize the goal successfully. Rather it talks about a complex whole: the learned or innate ability to cope with a network of feelings, perceptions, motives and emotions of tiredness and thirst, desire and purpose, in the presence of an apparent means of satisfying these. Sitting here concentrating on writing my lecture, I may not even notice the cup of coffee in my visual field until I come to the end of a paragraph, and become conscious of a bodily lack. Then a great many brain processes come into play: memory of what cups are and what steam indicates, calculation of how far, how big and in what direction the cup is, coordination of this with muscular movements of stretching and grasping, all monitored by visual clues in feedback loops that ensure that hand reaches handle, grasps, lifts, and, miracle that it all is, conveys cup to lip. This complex of processes is not even brought to consciousness unless, as Heidegger puts it in his homely metaphor, the hammer breaks, that is, until physical or neurological mishap or incapacity direct attention to what a miracle normal successful perception and purposive action is.

Several features of this model of perception/action processes should be noticed:

1. The simulations connecting inputs and outputs required for simple everyday functions have to be of great complexity. Arbib's diagram of a control program for grasping an object requires five kinds of inputs, eight "black boxes," and at least six feedback loops. Philosophers of perception have never taken sufficient account of the sheer fact of such complexity.

2. The account of perception and action is an account of "know-how," not "know-that," and hence is non-linguistic, not involving true or false propositions. Know-how is a category recognized by philosophers, but it is generally assumed to be reducible to know-that, and inferior to a linguistic expression of knowledge in propositions.

3.   The process of experience/perception/action is circular; inner and outer experiences trigger a perception and this triggers identification of a concept and perhaps an appropriate movement, but there is continued monitoring back and forth between inner and outer experiences and the behavior of the world.

4.   In most situations which are mediated by perception, movements of touching, grasping, manipulation, or even just moving the head, are essential ingredients in knowledge.

5.   The most important point to arise, however, is that nothing in this account impugns the reality of the external world and our "objective" knowledge of it. There are certainly elements of construction in the account: we rely on the presence in our minds of schemas, which may be innate or learned. They provide us with hypotheses about what and where an object is, and with habits for carrying out our purposes with regard to it. It is not even necessary for our communication with each other that the schemas be identical in different people's heads--there are for example many ways of conceiving a map of the Twin Cities that will enable us to move around in them successfully. That we are in contact with reality through our different schemas is ensured by the fact that we may make maps.

Words therefore lose information every time they reduce knowledge or experience to language. Moreover in practice our language cannot satisfy either of the crucial requirements of a deductive logic: it is never completely consistent, and it is never completely unambiguous or exact in meaning. The continual circular monitoring and adjustment of schemahypotheses and experience makes consistency a relatively unimportant condition--we often hold inconsistent hypotheses together and oscillate between them. What matters is not consistency but that we have a sufficiently approximate fit between the world, our experience and our actions to pursue our purposes successfully and correct mistakes as soon as possible.

It may be objected that, even if all this is correct, nevertheless inconsistencies, ambiguities and inexactness can be tidied up in advanced science, when we have discovered the right concepts and right laws. It may be argued (as for instance by Quine) that while the essence of nature is not visible to immediate perception, nevertheless science is that highly successful methodology that eventually discovers the essences and true causes, and can therefore be expressed in consistent, unambiguous and exact, i.e., propositional, language, and deductive logic. But notice what must be metaphysically true about nature for this to be possible. Nature must embody, as Aristotle thought, essences or fixed species of things, each of which can be given its correct "name" or definition, and the behavior of things must be regulated by universal causal laws which enable us to extrapolate the meanings of words to infinity, along with predictions of law-like behavior.

There is no sufficient reason to suppose that this metaphysics of nature is correct. Nothing in the use of language or the practice of science requires us to believe that the world is structured by essences and universal causes in this way. Science is successful only because there are sufficient local and particular regularities between things in space-geology or cosmology for example, but it is elementary mathematics that there is an infinite gap between the largest conceivable finite number and infinity. The accurate applicability of propositional language, with its strong requirements on linguistic meaning and truth, has become a metaphysical prejudice which has prevented recent philosophy of science from exploiting its radical implications for the nature of knowledge in general.

Once we have abandoned the Aristotelian ideal of true theories corresponding with the essences of the world, we are free to interpret the "end" or goal of science in another way. This goal is, I shall argue, <u>exactly the same</u> as that of everyday perception/action processes. It is to manipulate the world successfully for our purposes, and to discover local regularities in order to make sufficiently accurate predictions and hence to cope with and as far as possible control our environment. It is true that this goal is similar to that of all animals who require adaptive habits in order to survive, but it does not follow that this is not the root of science also. Science differs in its development of sophisticated mathematical and experimental techniques, but it need not therefore differ in its overriding human purpose.

Even the most elementary theories in early modern science exhibit the features that we have found characteristic of perception/action cycles. The immediate unthinking local observation that the earth is flat and stable under our feet is corrected by the Copernican-Galilean theory by taking account of a wider domain of evidence from the heavens. This theory commits what Galileo called a "violence upon the senses," and in his dialogues it is the pure empiricism of his character Simplicio (the Aristotelian) that has to be modified by a circular process of perception-theory-experimental-test. The theory commits violence upon immediate experience, but can itself explain and reinterpret this experience, and it is then perfectly congruent with it and with a wide variety of other observations of sun, planets, moon and falling bodies.

Circular and predictive enterprises of this sort form the stock-in-trade of modern science. Both their relative success and, in modern technology, their increasingly obvious failures, witness to the objectivity of the applicable parts of science. By an extension of the circular feedback method of ordinary learning, we have discovered many regularities of local phenomena, and we have many hypotheses, models and networks of schemas in scientific heads and in the journals and textbooks. This process of learning continues in cycles of interaction and modification with experience.

Is this what is meant by the end or goal of science? Many philosophers and some scientists will reject it as crude instrumentalism, or as an obsession with "mere" engineering. They will protest that it is not the know-how of technology that science is valued for, but the beautiful order and regularity of the world underlying its phenomenal flux whose discovery is the proper end of science. We seek true theories, they will say, not power and control, which have in any case brought many aspects of our civilization to the brink of disaster. By speaking of prediction and control, however, I do not necessarily mean big technology. What is required for what we call science is the possibility of empirical test of some sort--whether simple observation, as in the case of the more obvious motions of the heavenly bodies, or laboratory experiment which tests laws and theories about, for example, unobservable micro-particles. Big technology may be a spin-off in both cases, but it is not necessary to the conception of pure science as discovery of local regularities and successful prediction from them.

There are two sorts of response to the realist's protest that science is primarily concerned with true theories, not empirical success. We may agree with it, and try to rehabilitate the goal of theoretical discovery. To do this we would have to overcome all the problems about this conception of knowledge that have arisen in recent history, philosophy and sociology of science, and which form the agenda for this Conference. Maybe all these difficulties can be overcome, and sense be made of the ideal of closer and closer approximation to the truth expressed in propositional theories. Attempts to do so have not so far been very successful. This is not for want of trying, because the "relativist" moves of the 1960s and '70s by Kuhn, Feyerabend, and the rest have produced a realist backlash among philosophers anxious to re-establish what they call "rationality," which often means the unique pretensions of theoretical science to truth. This backlash has been all the more strongly motivated by the apparent absence of other serious accounts of rational and justifiable knowledge.

The "end" of this conception of science suggests approaching the question in a different way. Perhaps the whole imperialist claim of theoretical science to be the royal and single road to knowledge has been a profound mistake. Perhaps we should be looking in another direction for an account for knowledge, starting with the elementary facts about perception and mental construction made familiar by AI and cognitive science. As we have seen, these do not in the least impugn the objectivity of our perceptual knowledge; neither do the parallel accounts of scientific method and goals impugn its objectivity. But this is objectivity of a more humble kind than that sought by the theoretical realists. It is pragmatic, approximate and local in application rather than causally universal, its linguistic expressions are often ambiguous and metaphorical in meaning, its arguments are circular, inductive, analogical, rather than a priori and deductive. Theories have their very important place in this account of science, but they are like the schemas of perception theory, part of the circle of hypothesis and action, always modifiable and sometimes rejected by the monitoring function of action and application.

Ironically, such a view releases theories from the straight-jacket of unattainable propositional truth, and enables them to be seen as much more humane ingredients of science than the realists have contemplated. They are indeed human constructions, arising spontaneously in the history of ideas from the whole intellectual, social and technological environment of human thinking. Atomism, for example, is a perennial fundamental theory that owed its origin to observations of the behavior of air, water, and heavy particles, as well as to man-made machines and the discovery of the irrational in mathematics. It always carried reductive implications: nature is nothing but mindless matter in motion. Such a mechanist cosmology has always had metaphysical and ethical consequences for human life--the soul is a thin gas, the vicissitudes of life are either rigidly predetermined or are nothing but chance, the questions how we should live, what is the good life, are pushed to the sidelines, and their answers cannot be other than arbitrary relative to the backdrop of the vast play of atoms signifying nothing.

We owe to recent studies the insight that scientific theory is just one of the ways in which human beings have sought to make sense of their world by constructing schemas, models, metaphors and myths. Scientific theory is the particular kind of myth that answers to our practical purposes with regard to nature. That theory also often functions as persuasive rhetoric for moral and political purposes is not part of this pragmatic function, but it shares this character with no less important functions of other social myths.

The institution of science has made the single exception of its goal of "objective," empirically testable knowledge. This objectivity is defined as belonging to successful empirical outcomes, "facts" are what can be treated as objective and repeatable in those outcomes, and "truth" is a property of the description of those outcomes. Insofar as theories participate in objectivity, fact and truth, it is in virtue of their place (when successful) in the perception-prediction-test-correction cycle, not in virtue of an independent access to a factual world beyond the empirical.

Science defines its own values as a consequence of what may be called a collective social "decision" within all the institutions of science, since roughly the 17th century, that intersubjective experimental test is going to be the sole criterion of adequate factual knowledge. At a stroke transobservational and transnatural "facts" were defined out of existence insofar as they cannot be confirmed by the pragmatic criterion of successful test. From this point of view the realist interpretation of theories is the attempt to borrow the clothes of pragmatic science to cover the ontological nakedness of theory. More than this, realism of theory has become a mythical self-image of science. Because discovery of mere local phenomenal regularities seems a poor reward for pursuit of what was claimed to be the unique road to knowledge, and because all other traditional roads became intellectually discredited, it has become a psychological necessity to look to scientific theory to answer all those deep questions that have ever exercised the human race: What are the origin and destiny of the universe?, of human life?, What is the human mind or soul?, What is the good society?, How shall we live? The sciences, as so many cosmological myths, have been called upon to give us this framework in modern society--notice how they are taught in schools, and popularly presented by the media. But this mythological fare is impoverished and superficial when compared with traditional social and religious myths, as can be seen from constant attempts to relate the sciences to "human values" or derive an ethics modern science lay precisely in the rejection of all value implications except that of pragmatic success, and pragmatic success depends on such rejection, because the natural world is not built to answer to our values. At the time of the scientific revolution social requirements were satisfied, not by science, but by a generally accepted ethics and theology.

Misinterpretation of science to provide cosmological myths prompts the question: can we rehabilitate more traditional types of "knowledge" which have performed the role of cosmological myth better than the sciences, and which must now be seen as serious rivals to the myths drawn from science? This is a task that requires nothing less than a reinterpretation of the concepts of knowledge, objectivity, truth and facts. It is not a task to be accomplished quickly, perhaps not in this generation, but we can perhaps develop some hints from the analysis of scientific knowledge itself, once its pretensions to be uniquely objective are exposed.

The basic concepts for an alternative view of scientific theory are those of model and metaphor. Like mental schemas of perception and action, they describe a structure and tell a story about the natural world, for example atomism, within a particular set of categories and presuppositions. These are subject to radical historical and social change, while in the case of science always being constrained by pragmatic tests. But now we want to emphasize the features of scientific theory which transcend pragmatic success and give us, as do other myths and world-models, a means of structuring and understanding our world.

A new theory of knowledge requires a new theory of language. This must break with what the historian of the Royal Society in the 1660s called a "close, naked, natural way of speaking," and must be recognized to be inexact and local in application, and to reply on metaphor, analogy and model rather than univocal meanings and logical deductions. Scientific models are networks of systematic metaphors which build upon the essentially metaphoric character of natural language. The tests must be social and psychological, and as various as human needs and purposes. They include coherence with accepted systems of morality and with universal natural needs. They endow human life with a framework, meaning and goal, a beginning and an end. They answer to the perennial dissatisfaction with things as they are, and the longing for salvation. They satisfy the need for prayer and worship of something other than the world. They create and mould companies of believers. How these and other criteria of social truth function is the task of the human sciences, ethics and apologetic theology to discover.

Is this not a circular argument, and a craven surrender to the worst form of relativism? It seems to say that the true and the good are what we think they are, and answer to no reality beyond the natural and social. The approach is certainly relativist, but it is possible to be relativist, or better, pluralist, with regard to knowledge, but absolutist with regard to reality. Relativism with respect to knowledge is an inescapable consequence of our sensitivity in this century to the unconscious forces shaping the individual and to the power of social institutions to structure beliefs. These are the sensitivities that spell the end of the traditional conception of science: we are all to some degree Freudians and Marxists now. But as we have seen in the case of science, relativism refers to knowledge constructions, not necessarily to reality. There is an inevitable pluralism of theories, and even schemas for normal perception do not have to be the same for every perceiver. Theories and schemas are relative to human needs and to culture and history. The real, however, is that which objectively underpins our cultures and constrains our purposes, whether it be called a natural, psychological, moral, aesthetic or religious reality. The plurality of theory-systems is not subjective or arbitrary: they are thrown up by our various cultures, and are like Kuhn's normal science, to be worked with critically until something more acceptable to experience and consensus comes along.

There is indeed no Archimedean point for knowledge to stand on, but if we believe in transnatural as well as natural reality, we may trust it to constrain our social and psychological belief systems, as nature constrains scientific theories. These constraints may operate only in the very long run, and may not even be progressively more successful throughout history. But the collapse of a rationalist theory of knowledge means that in the end there is nothing to choose between on the one hand, belief in God, or a realm of values, or some transnatural reality that is a surrogate for these, on the one hand, or on the other hand, the kind of relativism described in the introductory literature for this Conference as "what we know is a function of time, place and the accidents of communities and their conversation." It may be that the latter belief is not even socially viable, in which case relativism itself only becomes a real issue in special and local historical circumstances, such as those which are perhaps now coming to an end. This is one way in which the empirical and transempirical constraints work in our own time.

QUESTION

HACKING: I would like to ask a rather materialist question which would be I suppose inevitable in the technological component of our meetings. I am very sensitive and inclined more than many of my philosophical colleagues to agree with Mary Hesse that theories are to be thought of more in terms of models and even metaphors than has commonly been the case. I am also very sensitive to her comparisons with myth. But I'm worried about the things that these are models for. If we take an example which Glashow mentioned of a theoretical entity, the neutrino. There is a project which will probably go ahead, the building of an enormous neutrino detector, in which neutrinos will be used as energy sources which make high energy physics look silly. We will be using neutrinos to mess around with other parts of the world. Now I find it difficult to fit this realist way of speaking with a conception of theories as models going off to analogies, going off to myths. I'm happy for you to say, well, all our pictures of the neutrino are just pictures, just our models, but I'm inclined to say that the neutrinos are there and that is different from any of the great myths which have moved us in the past. And that just makes me want to hold back on the use of the word myth here.

HESSE: Let me ask the question: Is phlogistin real? Because that's kind of neutral. That's a long way back, and we can perhaps see the essentials, what the questions are. It seems to me the answer is Yes and No. The answer is yes because Priestly did some quite elaborate experiments. He combined various substances. He poured sulfuric acid on zinc, and something was emitted which was inflammable. One could ask questions about its weight. One got different answers. There were things that were happening which could surprise Joseph Priestly and would constrain the things he wanted to say about it. It so happens that we now having reclassified the chemical elements call it hydrogen. Now this is not to say that there wasn't something that he contacted which was real and we call it hydrogen. It's now very comprehensively linked in with all of our other theories. It's something which we don't see. But whatever future theories come in chemistry, there is going to be something called hydrogen. Although, notice it's now become two or more different substances. So in that sense when Priestly said this phlogistin was being emitted, he was speaking the truth. Phlogistin existed in that example. Another element in phlogistin does not exist because he had a theory in which he identified this substance with what was going on in burning, in rusting, and actions which in our reclassification we think are concerned with oxygen, a quite different substance. He had a theory in which all of this was very coherent. It wasn't for its time a bad theory. Phlogistin was a name for the phenomena in all these diverse experiments. Our current knowledge of course says that hydrogen is not something which is equivalent to oxygen or negative oxygen, so phlogistin is not something. I think similar answers ought to be given about the existence of neutrinos. The people who resist this say, ah but Priestly lived back in the 18th century. We're much more clever, we have physical theory which is going to stay, with a few modifications. We're not going to have radical modifications? How do we know? I think every successful theory in the past has been very self-centered and has seen its successes as being permanent in a much more comprehensive way than they were justified in doing. I see no evidence why our theory shouldn't be similar, probably everything positive we are now saying about neutrinos is right; it's only that in the future they will be described within a different theory, perhaps even radically different.

ELVEE: A question from the floor. In the sense that Nietzsche intended his aphorism, God is dead, would you say "science never lived." In other words modern science never could be expected to provide moral foundations, other foundations for western culture?

HESSE: I think this is right, and the two statements are not unconnected, because when God was not dead, there was something which could hide the fact of the imperialist pretensions of science. The emperor had no clothes because the clothes were being provided from elsewhere, namely, the belief in God. So, I think this is right. Because the 17th century, in making what I call their decision to pursue experimental knowledge, did not in general, have imperialist pretensions. Many of them knew perfectly well that they were excluding all value in science, and so they were doing this for the purposes of science itself and that values were found elsewhere. There were two books--one was written by God in nature, one was written by God in the scriptures and in the human heart, which provided the other set of values for people to live by, and most of them, Newton included would have gone on to say that the second locus was much more important for human lives.

How to Think About the End of Science

GERALD HOLTON

* We have been brought together to discuss whether science, traditionally the continuing source of new insights, of material progress, and of intellectual emancipation may now be coming to a close--to its <u>end</u>, not merely to the recognition of the limits on the power of science, limits of which scientists themselves on the whole are quite aware. In fact, in my research I find the better the scientists the more likely they regard themselves as failures, because the best scientists are most impressed by the limits of their understanding. They measure themselves from the top down, not, like ordinary mortals, from the ground up. But when faced with the topic of this Conference, "The End of Science?", any scientist is likely to wonder why the notion of the decline of science as we know it is being raised at this time, when most of the branches of science are more powerful than ever, and the discoveries ever more remarkable. And what would it be like in a world in which scientific research has ground to a halt because the necessary intellectual support no longer exists there? In a few minutes I shall look briefly at the essentially emotional base for the proposed death sentence. But my main task is to explore the intellectual component of it, and let me begin with that.

I

To the historian of science, the notion of the decay and death of science is neither a contradiction nor a novelty. The idea has been proposed many times in the past. To give just one example, toward the end of the 19th century, a number of new problems could not be solved by the then-current mechanistically based physics. In disappointment, the European scientist Emil Du Bois-Reymond wrote that there science had at last come up against an unbreakable barrier of understanding, beyond which we shall always remain ignorant. The cry was "Ignorabimus," and soon it was converted into the more exciting slogan, "the bankruptcy of science." It spread quickly, urged on by some philosophers of science who made the entirely inappropriate demand that scientists should be able to discover through their research the ultimate metaphysical reality behind phenomena. The whole epidemic collapsed when the presumably bankrupt science suddenly gave rise to such advances as quantum theory and relativity.

However, we can count on the persistent recurrence of this fascination, and therefore our task today seems to me to learn how best to think about this topic as a whole, how to think about the possibility of an eventual end to science. And here history will help us. For with very few exceptions, virtually all proposals to this effect are driven by just one or the other of two fundamental thematic ideas, or if you will, by just two metaphors.

One of these represents science as advancing essentially along a meandering, but on the whole rising, line; it recognizes the existence of occasional plateaus, even temporary downturns, but also sprints of exponential growth, and so on the average there is a more or less steady increase in the state of scientific knowledge. The other view is that of scientific activity rising and falling in a cyclical manner. The adherents of the first view, which I shall call the "linearists," tend to come out of the background of having actually done research in natural science. They tend to see science as largely an autonomous activity, not primarily driven by external forces. A typical image that emerges from their writing is science as an advancing river system, parting into branches and combining again, as it makes its way toward one great oceanic unity, if you will a holistic understanding of the natural world.

The "cyclicists," on the other hand, tend to think of science not as a goal-directed, progressive, cumulating activity. They are apt to have the image of a biological organism, based on the metaphor of the cycle--run through once or repeatedly--from childhood and youth to age and death, or the closely related political metaphor of periods of revolution, followed by a normal state, followed by yet another revolution, leading to yet another incommensurable state--a sequence of paroxysms or changes of mind that leave no hope for certifiable progress. These cyclicists come more often from among social scientists and historians, and, contrary to the linearists, they see science to be significantly driven by social processes. In the extreme case, they think of science as just one expression of some general spirit of the time--a by-product of it, as it were--or even chiefly a matter of "social construction," not essentially different from chess.

As is generally the case with thematically opposed positions, one cannot expect to decide for one and against the other by some simple test. Instead, I shall try today to set forth these two scenarios for science, by presenting in turn the arguments of the one most eloquent proponents for each. This encounter with two interesting minds will at the very least help us to understand better what to do with this question at the intellectual level.

II

Before I get captivated by either one of these, I want to pay the brief visit I promised earlier to the more emotionally based motivation behind this periodic call for a stop to science. While they are unlikely ever to be taken fully seriously, they can create a climate of confusion and mistrust. A few minutes ago I asked you to consider what life would be like if continued research really were to cease. Let us be quite clear about it: Under such conditions, mankind would not simply settle smoothly into some Polynesian paradise existence, or return to an agrarian Eden. Instead, mankind would be facing almost unimaginable catastrophes. To put it plainly, our planet is not in equilibrium; we have destabilized it with our ignorant meddling, and current knowledge is insufficient to assure a sustainable future. Life in the 21st century, when the students here are supposed to reach the peak of their career, will not be enviable and may not be bearable without a great deal more scientific knowledge than we now have. In short, the topic we are dealing with here is far too serious to permit the comfort of mere academic rhetoric, or the seductive chatter about the so-called Re-enchantment or Tao of science, or escapisms into the crypto-sciences.

The animating force for all these flights from reason is not really a belief that science somehow has lost its epistemological base and warrants. It is rather, at best a fear, deep down, that the ever more rapid sequence of scientific advances, and the emancipating effects following from them from Copernicus on, have brushed aside as superstitions some of the instinctive bases of self-confidence for much of the population, and at the same time have increased through technology the potential for damage which our violent instincts can cause to ourselves. These are very valid concerns, and they are being seriously addressed by scholars and scientists. I hope this will be done even more vigorously in the future.

III

Having touched briefly on the psychological background for today's topic, let me return to the two main choices for its intellectual base. In my judgment the best representative that I can bring to you of the cyclicist school of thought of the fate of science is what remains to this day one of the most fascinating and outrageous books, a book completed after ten years of labors by an obscure and impoverished German high school teacher, then in his thirties, with a doctorate in Greek mathematics and an encyclopedic ambition. His 1200 pages, much of it written by candlelight during the first World War, offered a grand Teutonic theory of both the past and the future course of history, interspersed with dramatic predictions, a good share of absurd-sounding speculations, and some shrewd insights. But the arresting overall conclusion of his book was revealed even by its original title, Der Untergang des Abendlandes, the sinking away, the annihilation of all Western civilization, including its science. The subsequent English translation gave the title of the book only inadequately as The Decline of the West. The author's name was of course Oswald Spengler.

His enigmatic work, published in July 1918, just as the terrible war was grinding to its bitter ending, was an immediate sensation, an irresistible challenge. The debate about it, in which scientists also joined, continued for decades. In his critical study of Spengler, the historian H. Stuart Hughes observed that despite all its shortcomings, and even because of them, "the book remains one of the major works of our century, the nearest thing we have to a key to our time." And indeed, as you will see, in it you will find in stark and extreme language precursors of today's arguments, familiar from the writings of Arnold J. Toynbee, Spengler's direct successor and from the works of Theodore Roszak, Charles Reich, the last books of Lewis Mumford, the so-called "New-Age" authors, and even some of the writers on radical feminist science.

Spengler's key conception is that for every part of mankind, in every epoch, history has taken fundamentally the same course. And from that inevitable course follow in each case naturally the specific forms of activity, whether social, political, literary, artistic, spiritual-religious, or indeed scientific. Each of the mighty cultures of mankind--for example the ancient Indian, Chinese, Arabian, and the classical Greco-Roman--was not only as valid and significant as is our own Western civilization, but each is a drama with analogous structure. That is, each goes through the same season-like cycle, from its own nascent spring to its eventual burial in its own winter. Thus our own inevitable destiny in the West is to go to dust according to a timetable that can be calculated from the available precedents. Our time, Spengler said, corresponds not to that of Athens in the time of Pericles, but to that of Rome under the brutal Caesars. We are very near the end of our cycle. Of great painting, music, architecture or science, there can be for us no longer any hope. Our best strategy, he says, is to be bravely resigned and try at least to get a first glimpse of the rise of the next wave, which is coming from the East to triumph over the West.

Spengler tells us how each cycle progresses, from start to finish. Following Nietzsche, Spengler declares that each beginning is characterized by what he calls the Apollinian spirit, symbolized by the sensuous, individual body which we can see in classical Greek sculpture. With it goes a world view embracing attention to form and the organic, rather than to the mechanical or mathematical interpretation of experience that took its place later. It is the time of contemplation, not yet of investigation, of faith rather than skepticism, of high art rather than what he calls merely the "cult of science."

At some point into this cycle, however, there occurs a kind of historic change of phase of the Apollinian soul and of the culture which it animates. It gives way to its opposite a so-called Faustian one, which starts with a rather Germanic form of lonely romanticism, a yearning for the infinite, but gradually becomes more and more intellectualized. Thereby what was a culture is changed into a mere civilization. What then counts is the notion of causality instead of destiny; attention to cause and effect rather than what Goethe had called "living nature"; to abstractions such as infinite and empty space, rather than the palpable earth which you can feel and smell. In a civilization, the primacy of soul is replaced by intellect; concern for human needs degenerates into debates about money; mathematics pervades more and more activities; the principle of causality is forced on the understanding of phenomena, and nature is reinterpreted as a network of laws within the corpus of "scientific irreligion."

This transition from culture to civilization was completed in the 4th century for the world of antiquity in Europe; and Spengler proposes the same transition began in the late 19th century for the cycle of our Western society. As in past cycles, the phase in which we find ourselves will not end abruptly. It will linger on for some time. Writing over 70 years ago, Spengler declares that we have entered the last stage in world politics, too, which is the replacement of "the idea of the state's service" by the naked "will to power." As Nietzsche had predicted, ours would be the century of tyrants, of weapon-hungry Caesars engaged in a struggle for world rule--even as an entirely new culture is getting ready to take over the field.

IV

It is of special interest to us that in Spengler's somber drama, science plays a crucial role. The Faustian element in science, Spengler informs us, is exemplified succinctly by the famous confession of the physicist Hermann von Helmholtz, who wrote that "the final aim of natural science is to discover the motions underlying all alterations, and the motive forces thereof; that is, to dissolve all natural science into mechanics." This urge is not merely an expression of the universal longing to find the One in the Many. More specifically, Spengler notes, in our science "the seen picture of nature [is converted] into the imagined picture of a single, numerically and structurally measurable order." If he were writing today, Spengler would perhaps have replaced Helmholtz's quotation with the recent one by the physicist Leon

Lederman, who, encouraged by the current success of the unification program in physics, has mused that the aim of science now is to reduce all natural phenomena to one equation that will fit on a T-shirt.

Now Spengler introduces his most startling idea, one that has become familiar in new garb also. He warns that it is characteristic of the winter phase of civilization that precisely when high science is most fruitful within its own sphere, the seeds of its own undoing begin to sprout. This is for two reasons: the authority of science fails both within and beyond its disciplinary limits, and an antithetical, self-destructive element arises inside the body of science itself that eventually will devour it.

The failure of science's authority outside its laboratories, he says, is due in good part because of the tendency to overreach and misapply to the cosmos of history the thinking techniques that are appropriate only to the cosmos of nature. Spengler holds that the thought style of scientific analysis, namely "reason and cognition," fail in areas where one really needs the "habits of intuitive perception" of the sort he identifies with the Apollinian soul and the philosophy of Goethe. And even in the cosmos of nature there is an attack on the authority of science, arising from within its own empire. For every conception, even in science, is at bottom "anthropomorphic," and each culture incorporates this burden in the key conceptions and tests of its own science, which thereby become culturally conditioned illusions.

For example, Spengler goes on, "to the Classical belonged the conception of form; to the Arabian, the idea of substances with visible or secret attributes; to the Faustian--ours--the ideas of force and mass." In particular, the Faustian physics of the last 300 years has been a physics of dynamics and of "methodical experiments," both of which, Spengler says, are exemplifications of the will to power that imbues the civilization phase of a people, when "Nature is not merely asked or persuaded, but forced." All our rushing after positive scientific achievements in our century only hides the fact, he thinks, that as in Classical times, science is once more destined to "fall on its own sword," and so make way for the coming world outlook, what he calls the "second religiousness." Indeed, guided by his theory of cycles, he tells us "it is possible to foresee the date when Western scientific thought shall have reached the limits of its evolution." And in one of the handy chronological charts which Spengler put at the end of his book, he allows us to find that date. It is the year 2000.

Indeed, to Spengler's eyes the signs of decay, disintegration, Untergang of science were clear already by 1918. Physics, he says--and note how familiar this has become--physics has been infected by an "annihilating doubt," as shown by "the rapidly increasing use of statistical methods, which aim only at the probability of results and forego in advance the absolute scientific exactitude that was a creed to the hopeful earlier generation." The possibility of a self-contained, self-consistent mechanics has to be given up because "the living person of the knower methodically intrudes into the inorganic form world of the known." Moreover, the ruthlessly cynical hypothesis, as he calls it, of the relativity theory strikes at the very heart of dynamics. The quantum ideas are equally destructive. And Spengler adds that he is alarmed "how rapidly card houses of hypotheses are run up nowadays, every contradiction being immediately covered up by a new hurried hypothesis." So, giving up the search for exactitude and absolutes, and adopting probabilism, have undermined Faustian science from the inside. Our inability, for example, to specify which atom in a sample of radioactive material will decay next points directly to the Achilles heel of modern science. It is as if the idea of destiny instead of causality has been unwittingly reintroduced into the nature picture.

And yet another, final cause for the self-destruction of the modern scientific world picture arises, he says, from its tendency to theory and to symbol orientation. For what is happening is that all the separate sciences are converging into one, a "fusion characterized by the reduction to "a few grand formulas" in the winter of science. But ironically, just this has led us now back precisely to what is the first and simplest activity in the beginning of every new culture, and is always part of its primitive religious spirit: that is, the preoccupation with numerical regularities. Number is part of the earliest religious belief and ritual; number mysticism appears in every faith in such sacred concepts as the relation of microcosm to macrocosm, or in the building of prehistoric structures that served both for religious rites and for astronomy.

All these internal cancers will shortly kill science as we know it, and we shall rediscover that at bottom mankind as a whole, he says, has never wanted to analyze and prove, but has only wanted to believe. What he calls this orgy of two centuries of exact sciences is ending, together with the rest of what was valuable in Western civilization. Indeed, the only activities which are on the ascent during this final act are economics, politics, and technology. And as a kind of postscript, in his later book, Man and Technics (1931), Spengler adds his opinion that advancing technology, with its mindlessly proliferating products, will also turn out to undermine the society of the West because, he predicts, there will be a failure of science and engineering education: the level of it in the metaphysically exhausted West will not be up to maintaining the advance there. The attraction of the scientific-technological profession is diminishing. "The Faustian thought begins to be satiated with machines . . . and it is precisely the strong and creative talents that are turning away from practical problems and sciences. . . . Every big entrepreneur has occasion to observe a falling-off in the intellectual qualities of his recruits." At the same time, the previously over-exploited races, "having caught up with their instructors," have begun to surpass them, and to "force a weapon against the heart of the Faustian Civilization." The non-Caucasian nations will adopt the technical arts and turn them against the Caucasian inventors.

V

Thus spoke the ancestor of our End-of-Science movement. It is obviously rather easy to find specific faults with this work, as it is for the repackaged versions that clamor for attention today. Let us leave aside that science shows no sign of coming to an end within the next ten years; our potential for mischief, together with the general lack of political imagination, do not rule out entirely that before the year 2000 science might perish together with mankind in an act of self-destruction. Let us also leave aside his dangerous celebration of anti-rationalism. But one cannot fail to note the frequent, basic misunderstanding about science by Spengler and his heirs. For example, the use of probability and of quantum causality is not an abandonment of all causality as such. The notion of entropy does not, as he thought, inevitably lead to the heat death of the universe. The subjectivity of the individual does not rob science of all claims to objectivity. And so on.

More seriously still, Spengler, who was really a 19th-century thinker, could not have foreseen the rapid internationalization of almost every aspect of science. Even if the Occident should in some deep sense eventually decay and some other culture takes its place, it is a safe bet that short of a return to total primitivism, their schools will also be teaching Newtonian dynamics and Einstein's space time and the Watson-Crick double helix. These wheels cannot be disinvented.

On the other side, I should point at least to Spengler's perceptive insistence that despite what he called the "irreligiousness" of science, there is a subterranean link between science and religion at their origins. And that particular, unpopular aspect of Spengler's cyclicist view had some analogy in the work of a very different person. It is in fact the person I have chosen to represent to us now the opposite, the linearist view of the fate of science. For this purpose, I could well have selected other scientists I have studied, such as Johannes Kepler or Hans Christian Oersted or Niels Bohr. But it will be more appropriate if I briefly analyze a remarkable three-page essay that appeared also in 1918, almost exactly when Spengler's book was published, an essay still representative of the linearist view today, written by a man of almost the same age as Spengler, and one who was then also still almost unknown outside his own circle. The essay was originally a speech given in honor of the 60th birthday of Max Planck, whose work Spengler had just found to be destructive to science. And the name of the young speaker, whose work Spengler had also singled out as a symbol of disintegration, was Albert Einstein.

He will be our guide for the second of the chief ways of thinking about the fate of science. Let us listen to him as he rises at that dark point in history in 1918 to give his short talk. In rendering his thoughts, I shall quote his actual words as often as possible.

VI

Einstein begins with an image, saying "The temple of science is a vast building with many different wings." In it, many are there who pursue science out of the joy of flexing their intellectual muscles, and others for short-term utilitarian ends. But happily, there are also a few who do it simply because of their deep longing for knowledge itself. What led those into the temple? They have two motives for doing science. One is negative--a desire to escape one's "everyday life with its painful harshness and wretched dreariness, and from the fetters of one's own shifting desires."

But there is also a positive motive. "Man seeks to form, in whatever manner is suitable, a simplified and lucid image of the world, a world picture," a coherent view of how the cosmos of experience hangs together, "and so to overcome the world of experience, by striving to replace it to some extent by that image. That is what painters do and poets and philosophers and natural scientists, all in their own way. And into this image and its formation each individual places his or her center of gravity of the emotional life, in order to attain the peace and serenity which cannot be found within the confines of swirling personal experience."

The picture of the world which the physicist is building is only one among all the other possible ones. "It demands rigorous precision in the description of relationships." Therefore, the physicist must be content with studying first an idealized world, where for example--as you know from your Physics 1--all friction is negligible. "This allows him to portray the simplest occurrences which can be made accessible to our experience." The more complex phenomena cannot be immediately attacked with the necessary degree of logical perfection and accuracy. Therefore, at the beginning of a problem we strive for "supreme purity and clarity, but at the cost of completeness."

But such simplifying reductionism--to which, let me add, the Romantic critics, from Goethe through Spengler to this day, are so opposed--this simplifying reductionism is only the first, preliminary stage in Einstein's theory of scientific advance. History has taught us, he continues, that once a world image has been achieved on the basis of simplification, it turns out to be at least in principle extensible to every natural phenomenon as it actually occurs, in all its complexity and its completeness. Reductionism is only a detour to the road leading to the eternal, synthetic laws.

And now Einstein reveals the long-range agenda for science as he sees it, the destiny of science: From the general laws "it should be possible to obtain by pure deduction the description, that is to say the theory, of _every_ natural process, including those of life." That program of the eventual unification of all exact knowledge of the kind to which natural science can aspire is the final aim, the _telos_ toward which Einstein sees science striving.

We may well note here that in the intervening years enormous progress has been made in this direction, for example by finding that a good deal of chemistry is just that part of atomic and molecular physics which really works; by discovering the bridge between biological and physical sciences via DNA; and by finding some deep links between behavioral aspects and one's genetic endowment or biochemical imbalance; and of course by the ongoing unification of the forces of physics. In short, in modern form the old theme of finding the One in the Many has become the stuff of which Nobel Prizes are made. It is no longer entirely the dream of our much-maligned Doctor Faustus, the Faust who in Goethe's drama exclaimed that either he would attain the knowledge of everything, or else he would have to remain a mere worm.

But to return to Einstein's talk: At this point he issues the warning that the general program for the eventual unification of all the sciences, while yielding ever deeper insights and being a powerful motivation, is likely to be one without any foreseeable end. The meandering line tracing out the advance of science is not terminating; we have an infinite task on our hands. One reason is that despite all our successes we really lack a reliable method or guaranteed algorithm, for we have to make do with the fallible capacities of human thinking. Far from embracing the stereotype of a relentless victory march of cold rationality, which in any case exists only in bad science textbooks, Einstein freely confesses here, as he was to do again and again later, and contrary to the then-reigning philosophy, that "to the [grand] elemental laws there leads no logical path, but only intuition."

Of course that does not mean that anything goes, or that science has lost its authority and is doomed to stumble blindly from one discovery to the next. While there is no logical bridge from experience to the basic principles of theory, and hence no proof of the validity of philosophical realism itself, in practice we have not only good tests for the degree of veracity of our theories. There is also the fact, the astonishing fact, that agreement is possible within our very heterogeneous scientific community. That is a sign that "the world of experience does uniquely define the theoretical system." Even though _a priori_ we had no right to expect any such correspondence, somehow the order we put into our theories can, and often remarkably does, turn out to correspond to the order others find in nature when they check our predictions.

Why is that possible? Why can our limited mind penetrate so often and so well behind the appearances to discern a few eternal laws? How can it find its way back and forth between the world of phenomena and the world of ideas? On that point, Einstein confesses freely, he has no certain answer. But that does not make him collapse in demoralized helplessness. He has a daring suggestion--that our minds are guided by "what Leibniz termed happily the 'pre-established harmony.'"

Gottfried Wilhelm Leibniz, the philosopher and contemporary of Newton, had postulated that our ability to discover the laws concerning material bodies is one aspect of the unity from which God created the two apparently separate entities of the universe, the spiritual and the material. Each of these obeys its own laws; but they can interact in sympathetic unison, somewhat in the way one stringed instrument goes into resonance and picks up the sounds made by a second one that is tuned to it. Or, to use Leibniz's own words to explain this possibility of a harmonious interaction, in which he uses an image that must have delighted Einstein: "The souls follow their laws . . . and the bodies follow theirs. . . . Nevertheless, these two beings of entirely different kind meet together and correspond to each other like two clocks perfectly regulated to the same time. It is this that I call the theory of pre-established harmony."

Scientists of our day are more likely to invoke an argument from the supposed evolutionary base of a correspondence between our ideas and our environment. They will do so less because of any proof, and more because they feel uncomfortable with the theological undertone of Einstein's metaphor, one which would have come more naturally to those who were familiar with Leibniz's discussion from their reading of the commentary on it in the writings of Immanuel Kant. But to Einstein just this undertone was by no means unwelcome or accidental. Having nearly reached the end of his essay with this image, Einstein returns briefly to the question of what motivates people to pursue science despite the lack of any guarantee of success or of a foreseeable end to the labor. It is wrong, he concludes, to trace this persistence "to extraordinary will power or discipline." Rather, "the state of feeling which makes one capable of such achievements is akin to that of the religious worshipper, or of one who is in love. [That is,] one's daily strivings arise from no deliberate decision or program, but out of immediate necessity."

VII

In the years that followed, Einstein continued on every occasion he could find to spell out and develop these views: Science is a program with an aim toward which one can advance, but one that has no foreseeable end. It is a mandate to produce the best objective description possible of the physical cosmos, while having to work only with one's subjective capacities and with essentially arbitrary concepts. It is an activity of persons able to combine logical rationality with intuition (contrary to the Spenglerian assumption of their incompatibility), who have the knack for acting both on hard evidence and on faith, and sometimes even on aesthetic grounds. Doing science requires analysis as well as synthesis. In short, science is the mobilization of the whole spectrum of our talents and longings, in the service of shaping more and more adequate world pictures. What to lesser minds looks like a mixture of mutually exclusive opposites between which one must make a choice, to Einstein seemed to be complementary necessities.

It is therefore not surprising that he, unlike Spengler and his followers, also saw no inherent conflict between science and religion, as Einstein hinted in this passing reference to the kinship between the scientist and the religious worshipper. In three additional short essays that also appeared in his volume Ideas and Opinions, he elaborated his deeply felt argument that scientific activity, the search for the evidence of rationality in the universe, is in essence a religious act. As you would expect, his description of what he called "Cosmic Religion" is not a product of sentimentality or of sectarianism; nor do religion and science, where they merge into Cosmic Religion, have much in common with the conceptions held dear by any religious establishment. Einstein's idea of God was not that of the biblical, intervening deity. Rather, his view, derived in part from Spinoza, serves as a necessary reminder that science, from its earliest beginnings to our time, has retained the signature of that single, undifferentiated totality which motivates our inherently endless human search both for explanation and for transcendence.

VIII

You will have noticed a few analogies between Einstein's linearist views and those of the cyclicists such as Spengler and his followers. For example, Einstein too was opposed to the more imperialistic claims of positivism. But the essential, overriding difference between them is that for Einstein, as for most modern scientists, the notion of an ending of science is a contradiction in terms. Today, also, there is not a shred of evidence to the contrary. In the absence of such evidence, we are being subjected to rhetorical diatribes of various degrees of eloquence, most of which have traceable antecedents in Spengler's pessimistic vision. For example, we hear that science is supposed to have the pretension to be the unique road to valid knowledge as such--whereas most responsible scientists, from Galileo to Einstein and Bohr to our day, have of course not claimed more than to be travelling on one road toward one particular and valuable kind of knowledge. We hear that science has failed by failing to deliver a proof for realism or other philosophical positions--whereas this has never been the purpose of science; or that it is now doomed to failure because it casts aside easily visualizable concepts in favor of abstract and mathematical ones (a charge familiar since the 17th century, but indicative chiefly of an underestimation of the flexibility of the mind of the next generation). And we hear of yet other newly fashionable fantasies--that science is merely the projection of Oedipal obsessions with such notions as force, energy, power or conflict; or that "faith in the progressiveness of scientific rationality" is so deeply and harmfully embedded in modern society that a "radical intellectual, moral, social and political revolution" is now called for; or that current scientific theories have inherently no more claim to be taken seriously than myths do.

All these are as invalid as were the claims of past enthusiasts of the opposite sort, who hoped that science would be the eventual guarantor of omnipotence over the affairs of mankind. All these only distract from the business of better understanding the power and limits of the various sciences, and of their application to the caring melioration of the lot of humanity. There is much to occupy us on that score. Far from being over-optimistic or smug, our best scientists, as I have noted earlier, are keenly aware how difficult their task is in the absence of agreed-upon methods, not to speak of guaranteed algorithms. On this point, the social scientist Otto Neurath proposed a memorable and apt analogy in his essay "Antispengler": "We are like sailors who on the open sea must reconstruct their ship but are never able to start afresh form the bottom. . . . They make use of some drifting timber of the old structure, to modify the skeleton and the hull of their vessel. But they cannot put the ship in dock in order to start from scratch. During their work they stay on the old structure and deal with heavy gales and thundering waves. . . . . That is our fate."

This picture of science as a self-constructing enterprise against great odds was further improved by the philosopher Hilary Putnam:

> My image is not of a single boat but of a fleet of boats. The people in each boat are trying to reconstruct their own boat without modifying it so much at any one time that the boat sinks. . . . In addition, people are passing supplies and tools from one boat to another and shouting advice and encouragement (or discouragement) to each other. Finally, people sometimes decide they do not like the boat they are in and move to a different boat altogether. And sometimes a boat sinks or is abandoned. It is all a bit chaotic; but since it is a fleet, no one is ever totally out of signalling distance from all the other boats. We are not trapped in individual solipsistic hells (or need not be), but invited to engage in a truly human dialogue, one which combines collectivity with individual responsibility.

In closing, let me invite you to consider the merit of the position which has emerged in response to the End-of-Science prophets. Let me suggest that it is the particular mission and talent of human beings, with only a precarious foothold on our endangered globe, to find certifiable truths with whatever limited means come to hand. We need not apologize at all for our innate, unquenchable impulse to seek rational meaning in those signals that reach us from beyond our frail frames. Nor do our innate impulses to seek transcendence have to be explained away or resisted, even in science. The challenge for our species, now more than ever, is to harness both these impulses, in the service of our creativity--and indeed of our survival.

Partial Bibliography*

Einstein, Albert, Ideas and Opinions (New York: Crown Publishers, Inc., 1954).
Hughes, H. Stuart, Oswald Spengler: A Critical Estimate (New York: Charles Scribner's Sons, 1952).
Schroeder, Manfred, Der Streit um Spengler: Kritik seiner Kritiker (Munich: C. H. Beck, 1922).
Schroeder, Manfred, Metaphysik des Unterganges (Munich: Leipzig Verlag, 1949).

---

* Note: I have generally gone back to the original German text and improved the English translations of quoted passages where necessary.

Spengler, Briefe. 1913-1936 (Munich: C. H. Beck, 1963). Translated as Letters
    of Oswald Spengler. 1913-1936 (New York: A. A. Knopf, 1966).
Spengler, Oswald, Der Mensch und die Technik: Beitrag zu einer Philosophie des
    Lebens (Munich: C. H. Beck, 1931). Translated as Man and Technics: A
    Contribution to a Philosophy of Life (New York: A. A. Knopf, Inc.,
    1932).
Spengler, Oswald, Der Untergang des Abendlandes: Umrisse einer Morphologie
    der Weltgeschichte. Volume I: Gestalt und Wirklichkeit (Vienna &
    Leipzig: Wilhelm Braumüller, 1918).
Spengler, Oswald, Der Untergang des Abendlandes: Umrisse einer Morphologie
    der Weltgeschichte (Munich: C. H. Beck, 1980). Contains, in revised
    edition, both Vol. I: Gestalt und Wirklichkeit, and Vol. II: Welthisorische
    Perspectiven originally published, 1922).
Spengler, Oswald, The Decline of the West (New York: A. A. Knopf, 1926
    (Vol. I) and 1928 (Vol. II)).

QUESTION

HACKING: I would like to ask a question about the linear picture and about science after 1918. It's commonly said for example that Kurt Gödel refused to publish the second half of a major result about Cantor's continuum hypotheses because it would only augment the anti-linear influences at loose in the world. And one can well see his more famous Incompleteness Theorem as an anti-linear influence much more complicated, but much more pervasive. This suggests that after Einstein's speech in 1918, there was a whole range of internal explosions within the linearity program which may make that way of thinking really obsolete. I wonder if you would like to comment on that?

HOLTON: There's no doubt that science goes through changes of phase and has done so. Think, for example of Galileo, in 1609, looking through his telescope and seeing this magnificent display of satellites around Jupiter and being warned by his colleagues not to take it too seriously, because as everyone knows, the glass distorts the phenomena and therefore he may be seeing an epiphenomenon. And now Voyager has traveled to the edge of the solar system and has checked out what Galileo saw, some three hundred years ago, and found that he was quite right, and the Inner Square law applies all the way out to Neptune. Yes, it is correct that between 1918 and 1989 there have been a number of explosions and we have the ability to be interested in the complex phenomena, rather than those which can be dealt with in the Einsteinian way. But even there, I think, the technique generally is to try to make a model with as little complexity as possible and then work toward the larger complexity. So I think I don't see much of a change between the Einstein of 1918 and our 1989 problems. If Einstein were to come back, or even if Galileo were to come back among us, you would find that they, with a little bit of instruction from our more clever graduate students, would get the melody very fast. That is to say, the general program can be understood by those people. But in between there have been details of amazing and great complexity elucidated, and that will always be true.

Cognitive Limits and the End of Science

GUNTHER S. STENT

1.   MACROHISTORY

The title of this Conference posed only one of several currently asked "The End of . .?"questions. Not only is it being asked whether science is coming to an end in our time but also whether music, and literature, and even history are as well. The theoretical foundation suitable for dealing with these questions belongs to a not-quite-respectable discipline styled *macrohistory*, which attempts to fathom the general, rather than the specific, dynamics of history. Hegel was its main founder and , as Gerald Holton reminded us, Oswald Spengler was its most notorious representative.

"The End of . . .?" questions usually turn on a particular macrohistorical notion, namely the Idea of Progress, of which British historian J. B. Bury[61] provided an incisive analysis earlier in this century.   The idea that history embodies progress was, for very good reasons, virtually unknown until the advent of the French Revolution. Once discovered, however, the Idea of Progress became the principal ideology of the industrializing nations. By way of an example of the kind of extrinsic influence on the course of science mentioned by several of my fellow panelists, the Idea of Progress provided the ideological infrastructure for the theory of organic evolution, as first put forward by Lamarck and as later improved on by Darwin. No sooner had the idea of past historical progress become widely accepted than it was claimed that progress was coming to an end. Thus the term "post-modernism" has now come into use to designate the condition of those fields of creative endeavor, especially in the arts--music, painting, architecture, literature and criticism--in which progress *has* ended in our time.

Many people, especially scientists, are not interested in "The End of Science?" palavers. They say, as some of my fellow panelists did, that they see no evidence that the progress of science is slowing down, that the very idea of an end of science is absurd, and that they just want to get on with their fascinating work. This is a view which I respect, even admire. But such people are not very well qualified for "The End of Science?" discussions: not being interested in the topic, they are usually unfamiliar with the standard arguments that have been debated for the past 150 years. So if they do take part in such discussions, against their better judgment, they keep on reinventing the (square) wheel.

There are other people, represented by at least one of the panelists, who may admit that science has contributed to human progress in the past, but find that science has now turned into a morally noxious enterprise. They regard contemporary science as a regressive, rather than progressive, social force conserving the unjust social structures of the past and fostering such evils as militarism, racism, sexism, and the destruction of the environment. These people tend to regard social responsibility as an integral part of science, not to be

---

[61]Bury, J. B. (1932). The Idea of Progress. MacMillan, New York. Reprinted by Dover Publications, New York, 1955.

divorced from its epistemological aspects. This view I respect as well. But these people, too, fall short of being good partners for "The End of Science?" discussions, since they tend to sermonize rather than argue. Their righteous homilies can do little more than keep the faith of the True Believers. The very existence of science-is-evil-and-must-be-reformed people is, of course, highly relevant for "The End of Science?" question since it provides an argument in support of the answer: "affirmative."[62]

Macrohistorians have identified factors both internal and external to science that they believe are likely to cause--or to have already caused--its end. These include the economic limits of science that arise from research becoming ever-more expensive and industrialized nations being unable to increase indefinitely the fraction of their gross national products spent on financing science. They include the physical limits, such as the maximum speed of light or the maximally practical kinetic energy to which elementary particles can be accelerated, that set barriers in principle to our study of the universe or of matter. They include the social and political limits that have recently arisen from anti-science movements in the industrialized nations, such as the Greens and the animal rights movement. They include the negative feedback loop between progress and the development of the kind of people who drive progress, that is of the Faustian personality type to which Oswald Spengler had drawn attention. They include Henry Adams' "Law of Acceleration,"[63] which purports to show that it is because of the autocatalytic or self-reinforcing, and hence exponentially increasing, rate of progress, that of all generations in the long history of our species we folks of the 20th century would happen to have the honor of being alive when progress, including progress in science, is finally coming to an end.

In my presentation, I will not address these factors, with which I have dealt in some of my previous publications.[64][65] Instead I intend to discuss three cognitive limits of science that have come into view in this century: a *semantic* limit, a *structural* limit, and a *subjective* limit.[66][67] Since the existence of these

---

[62]Stent, G. S. (1976). The new biology: Decline of the Baconian creed. Great Ideas Today, pp. 152-193.

[63]Adams, H. (1918). The Education of Henry Adams (Chapters 23 and 24). Massachusetts Historical Society, Boston.

[64]Stent, G. S. (1969). The Coming of the Golden Age: A View of the End of Progress. The Natural History Press, New York. 146 pp. (Paperback edition: 1971; Castilian edition: Seix Barral, Barcelona, 1973; Japanese edition: Misuzu Shobo, Tokyo, 1972; French edition: Fayard, Paris, 1973.)

[65]Stent, G.S. (1978). Paradoxes of Progress. W. H. Freeman & Co., San Francisco. 231 pp. (Spanish edition: Alhambra, Madrid, 1981).

[66]Stent, G. S. (1981). Ursprung, Grenzen und Zukunft der Naturwissenschaft. Jrb. Schweiz. Naturforsch. Gesells. (Wissenschaftl. Teil), pp. 64-72.

[67]Stent, G. S. (1983). Origins, limits and future of science. Rivista di Biologia 76: 549-579.

limits forms a barrier to an indefinite progress of our knowledge of nature, it is likely that the science of the future will be different from the science of the past.

2.   SCIENCE AND TRUTH

To begin my discussion, I shall try to clarify two concepts that are fundamental for our subject, namely science and truth. A thorough explication of these fundamental concepts would require more than the time available for my entire presentation, and so I will simply assert that science is an activity that seeks to discover and to make propositions about causal connections between perceived events in the outer world of things. It is by means of that activity that we try to order what would otherwise be a chaotic experience and to exercise power over the world. There are other activities that play a similar role, for instance religion and magic. Although these other activities are usually considered to be fundamentally different from science, to draw a sharp line of demarcation between what is science and what is not science is a philosophically very difficult problem, with which I do not wish to wrestle here. Thus I will not attempt to clarify in what sense astronomy is and astrology is not a science. What is essential is to note that science is a <u>semantic</u> activity, inasmuch as its purpose is to make communications about the world that have meaningful and significant, or at least interesting, content.

What does it mean to say that an alleged causal connection between events, that is to say, a scientific proposition, is true? Here, too, I cut short a very difficult philosophical problem and simply assert that a proposition is true insofar as it is in harmony with my internalized picture of the world, i.e. with my reality, and commands my assent. This view of the concept of truth is obviously not objective but subjective. It leads to the concept of an intersubjective truth as long as I am convinced that a proposition that is true for me would also command the assent of every other person qualified to make this judgment. But the ideal of an objective truth is reached only if God assents to the proposition.

How does a scientific proposition command assent? It provides a convincing answer to a 'why?' question about perceived events and it predicts events that actually happen. Sometimes an answer to a 'why?' question can lead also to an answer to a 'how?' question about the management of events. The power provided to us by an answer to a 'how?' question can be considered as a validation of the proposition from which that answer was derived. However, the validation of scientific propositions is another subject of heated dispute among philosophers. For instance, it is often argued that the truth of a proposition cannot be proven by a finite number of observations. Nevertheless, practicing scientists must believe in their heart of hearts that they can at least come close to validating their propositions: How could they spend their lives trying to understand the connections between the events of their experience if they were not convinced that their efforts would eventually produce truths?

3.   INTUITIVE CONCEPTS

After I have presented the scientist as a discoverer and communicator of the connection between events in the outer world of things, I now turn to the cognitive foundations of science, and thus to the question of how the mind is capable in the first place of posing and finding answers to 'why?' questions. Empiricists of the late 17th and early 18th century believed that the mind at birth is a clean state, on which there is gradually sketched a picture of the world, drawn from cumulative experience. This picture is orderly, or structured, because, thanks to the principle of inductive reasoning, we can recognize regular features of our experience and infer causal connections between events that repeatedly occur together. However, David Hume, one of the most brilliant exponents of empiricism, realized that the empiricist theory of knowledge suffers from a near-fatal logical flaw. As he noted, the validity of inductive reasoning can neither be demonstrated logically nor inferred from experience. Rather, inductive reasoning and the belief in the causal connection between events is evidently brought intuitively to experience.

Not long after Hume, Immanuel Kant showed that the empiricist doctrine of experience being the sole source of knowledge derives from an inadequate understanding of the mind. Kant pointed out that sensory impressions become experience, that is to say gain meaning, only after they are interpreted in terms of such *a priori*, intuitive concepts as time, space and object. Other intuitive concepts, such as causality, that is to say the connection between events via cause and effect, allow the mind to construct a picture of the world from that experience. And it is to this picture of the world that the concept of truth pertains, in that I have the intuitive belief that things really are as I imagine them (i.e., form an image of them).

But how is it possible that if, as set forth by Kant, we really do bring concepts such as time, space, object and causality to our sense impressions *a priori*, that these concepts happen to fit the world of things so well? In view of the great number of ill-conceived notions that we might have had prior to experience, it would seem nothing short of a miracle that our intuitive concepts just happen to be those that do fit the bill. But, as Friedrich Nietzsche first suggested a century ago and as Konrad Lorenz pointed out about half a century ago, this apparently miraculous concordance of intuition and the world can be easily explained by the theory of evolution. After all, if our brain is the product of natural selection acting on our distant ancestors, it would go without saying that the brain can possess hereditarily transmitted, that is to say, innate, knowledge of the world prior to any personal experience. Or as Lorenz put it, "experience has as little to do with the matching of the *a priori* with reality as does the matching of the fin structure of the fish with the properties of water."[68]

Even though these Kantian *a priori* concepts can thus be considered part of our innate biological endowment, they are nevertheless not 'inborn,' in the sense that they are already present in the brain at birth. Rather, as elucidated by Jean Piaget and his students, these concepts only arise in the course of a cognitive development in childhood. Piaget found that this development runs through a sequence of clearly recognizable stages, which are governed by the sensory-motor interaction of the child with its infantile environment. At first an infant does not ascribe constant size, or even identity, to the things of its surround and lacks the concept of an object. The concrete notion of an object that has identity and characteristic qualities appears only at a subsequent stage. Out of such concrete concepts there later develop the abstract linguistic, logical and mathematical

---

[68]Lorenz, K. A. (1941). Kant's Lehre vom Apriorischen im Lichte gegenwärtiger Biologie. Blaetter für Deutsche Philosophie 15:94-125. English version: Kant's Doctrine of the *a priori* in the Light of Contemporary Biology. In: General Systems (L. Bertalanffy and A. Rappaport, eds.) Soc. Gen. Systems Research, Ann Arbor, 1962.

modes of thought. For instance, Piaget found that the child must first develop the notion of invariance before being able to use words that refer to particular objects, or before having access to the concept of number. The abstract Kantian concepts of space and time appear in their mature form at yet later states.[69]

The significance of Piaget's discoveries lies, for the purpose of my discussion, in his empirical demonstration that our intuitive concepts arise during the childhood of every normal person, as a result of a hereditarily determined dialectic between the developing brain and the world of things. These concepts therefore represent biological givens, rather than contingent products of social or philosophical conventions. They are immanent properties of human reason, and to acquire them is what it means to grow up into a sane person. From this insight, it follows that humans were in possession of the intellectual equipment necessary for doing science at least since the appearance of *Homo sapiens* about a hundred-thousand years ago. But the first human achievements that we might recognize as being related to science came only about ten thousand years ago, in the late Stone Age--the successful breeding of domestic animals and crop plants, and the invention of metallurgy, pottery and brick-making.

Science, as I explicated it at the outset of my discussion, really got underway only about 2-1/2 thousand years ago, when the Greeks conceived of the idea that the world is governed by a limited number of natural laws open to discovery by the mind, from which answers to 'why?' questions about nature can be derived. Moreover, the Greeks considered natural phenomena as being independent of human emotions, and they placed humans as observers outside of nature, although they did not deny that humans are also subject to natural law. Thus the Greeks founded science as a search for objectively true propositions about the causal connection between natural events, i.e., for propositions to which God (or the gods) would assent. The meteoric rise of modern science began 400 years ago with Galileo's discovery that the natural laws conceived by the Greeks can be expressed mathematically. Galileo showed that it is possible to devise mathematical models, that is to say quantitative pictures, of the world that can give account of exactly measurable aspects of natural phenomena.

Let me summarize briefly what I have said. *Science is concerned with the discovery and communication of truths about the connection between perceived events. Scientists bring to this project their biological endowment of intuitive concepts, with the aid of which their minds construct experience out of sensory impressions. From this experience, in turn, they draw an internalized picture of that world. The picture is 'scientific,' insofar as it is based on the postulate of natural laws whose truth is supposed to be objective, i.e., independent of the mind that discovers them. These conceptual and cognitive foundations are at the origin of science, but at the same time they are also responsible for the limits of science.* I shall now try to bring these limits into focus.

4.   COGNITIVE INCONSISTENCIES

By the end of the 19th century, the project of science launched by the Greeks had become tremendously successful. It turned out that nature is indeed accessible to the mind and its intuitive concepts. Thanks to the understanding thus obtained, humans had gained extensive mastery over the world of things.

---

[69]Piaget, J. and B. Inhelder (1969). The Psychology of the Child. Basic Books, New York.

The excellent service which the Greek doctrines had rendered to modern science and technology appeared to confirm their validity in an impressive manner. Further scientific progress has brought to light cognitive inconsistencies. It was Niels Bohr who recognized that these inconsistencies arise because, for making scientific propositions, we are dependent on the intuitive concepts embodied in everyday speech, in the language that we have developed for orienting ourselves in our environment and for organizing our communities.[70] The models that science offers by way of explanation of the world are therefore linguistic images that have been constructed out of words borrowed from everyday speech.

These images were satisfactory as long as they pertained only to phenomena of our everyday experience. But this situation began to change as soon as physics ventured into subatomic or into cosmic dimensions. Upon attempting to orient ourselves in realism of space and time a billion times smaller or larger than those of our direct experience, we encountered serious difficulties. For it turned out that linguistic images of phenomena belonging to domains far beyond the middle domain that we can directly experience contain contradictions, or lead to mutually incompatible pictures of the world. In order to resolve these contradictions, we must alter certain hidden presuppositions that are implicit in our intuitive linguistic concepts. Making such alterations has a grave consequence, however, in that the meaning that these concepts have in the scientific context is no longer consonant with our intuition.[71]

One of the first of such far-reaching alterations of concepts of everyday speech was made by Albert Einstein at the beginning of this century. In his development of the relativity theory, Einstein had recognized that the experimentally observed constancy of the speed of light cannot be reconciled with the intuitive concept of time. This inconsistency arises from the presupposition hidden in the intuitive concept of time that its flow is absolute. In order to account for the constant speed of light, Einstein removed the presupposition of absolute flow from the intuitive concept of time, and thereby put forward the counter-intuitive proposition that the time of occurrence of an event depends on the frame of reference of the observer. Accordingly, there does not exist just one time but many times, to each observer his own. Einstein also dissolved the fundamental conceptual independence of space and time, whose intuitive development in the mind of every child Piaget had discovered to be a biologically given, natural process.

In the 1920s, the development of quantum mechanics brought forth a further erosion of intuitive concepts. Heisenberg's 'Uncertainty Principle' of quantum mechanics showed that the unavoidable interaction between observer and observed sets an instrumental limit to the objectivity with which phenomena in the dimensional realm of atoms can be ascertained. Quantum mechanics also led to the conclusion that we are not only unable to measure the location and the momentum of electrons with infinite precision but that such microscopic particles *are not* in any definite place and *have no* definite momentum until either one or

---

[70]Bohr, Niels (1961). Atomic Physics and Human Knowledge. Science Editions, New York.

[71]Stent, G. S. (1988). Light and Life: Niels Bohr's Legacy to Contemporary Biology. In Niels Bohr: Physics and the World (H. Fischbach., T. Matsin, A. Oleson, eds.) Harwood, Chur. pp. 231-244.

the other of these complementary parameters is measured by means of mutually exclusive experimental arrangements set up by an observer. Accordingly, such particles no longer correspond to the intuitive concept of an object, which ought to be in only one place and ought to move in only one manner at any given time. The gravest violation of intuition by quantum mechanics is undoubtedly its claim that the dynamics of an electron are not subject to the cause-and-effect chains thanks to which the events of the everyday world of our direct experience are connected. Instead, events in the microscopic world of atoms are governed merely by probabilistic, indeterminate laws. Einstein could not bring himself to accept this view because he did not want to believe that 'God plays at dice.'[72]

In the thirty years that have passed since Bohr's death, nuclear physics has taken further steps towards the conceptual alienation of science from intuition. Whereas the properties ascribed to the elementary particles of quantum mechanics--electrons, protons and neutrons--were already no longer fully consonant with the intuitive concepts of space, time, object and causality, the technical sense of words such as mass, charge and spin, in terms of which these properties were described, had still enough in common with everyday meaning that semantic contradictions could be at least recognized. But with the advent of the quark theory of matter, chromodynamics, a mode of speech entered physics that, although making use of words of everyday language such as 'up,' 'down,' 'strange' and 'charm,' no longer employs these words in any metaphorical sense. Here 'up,' 'down,' 'strange' and 'charm' have nothing at all in common with their everyday meanings and cannot therefore evoke any picture of the world. Evidently the properties of quarks designated by these words are purely formal symbols, unconnected to intuitive concepts.

We may now ask to what extent contemporary nuclear physics is still within the realm of natural science. In what respect can one still designate as 'true' the propositions of the chromodynamics of quark particles? How can such propositions command my assent if they do not allow me to imagine anything that I can bring into harmony with my internalized picture of the world? Is the purpose of chromodynamic theories still the discovery and communication of lawful connections between events in the world of things, or is their purpose merely to predict the results of very high-powered experiments? In fact, is it even possible that the theories of chromodynamics extend our power over the world of things? That is to say, will they supply us with answers to practical 'how?' questions even if they do not lead to pictures that can be grasped and to meaningful replies to theoretical 'why?' questions? Evidently here we encounter one of the altogether central questions for the future of science.

In order to summarize briefly this part of my discussion, I once more recall that *science made a triumphal rise in the two-and-a-half thousand years since its launching in ancient Greece. Science showed that nature is accessible to the mind with its intuitive concepts and provided for us a very extensive power over the world of things. At the beginning of this century, however, further progress in physics suddenly brought to light cognitive inconsistencies that made it necessary to alter our intuitive concepts. The alterations that had been brought about by relativity theory and quantum mechanics brought science to a limit, which was subsequently transcended by chromodynamics. This limit is a*

---

[72]Stent, G. S. (1979). Does god play dice? The Sciences, March: 18-23.

'semantic' one, beyond which the world of things can no longer be understood by use of our intuitive concepts.. We may ask whether, once that limit has been transcended, science can still be considered to be an activity that seeks to discover and make propositions about lawful connections between events in the outer world of things, or whether it has become a mere formalism for the prediction of the results of experiments. Should it turn out that chromodynamics does lead to practical advances, just as the relativity theory and quantum mechanics opened up the atom as a practical source of energy, then, in my opinion at least, it would be a genuine miracle, one that could not be explained away by facile arguments about natural selection of *Homo sapiens'* brain.

5.   SECOND STAGE INDETERMINISM

So far, I have tried to show that science is approaching, or has even reached, a semantic limit. Now I shall discuss a second kind of limit, which similarly came into view in this century. This limit was first recognized, or at least clearly exposed, by the mathematician Benoit Mandelbrot in the 1960s, while he tried in vain to fathom the statistical fluctuations in the price of cotton.[73][74] The difficulties which Mandelbrot encountered in the course of that study caused him to develop an epistemological argument that accounts for his failure. The applicability of this argument however transcends economics, having general validity for the search for answers to 'why?' questions. To appreciate Mandelbrot's argument, we first note that science is a statistical endeavor. The scientist searches for a common denominator, or a structure, in the ensemble of events in which he is interested. As soon as he has recognized such a structure he infers that the vents are connected, and he tries to find a natural law that would account for the cause of this connection. Thus, for the discovery of causal connections, the scientist depends on many similar, or related, events. A unique event, or at least that aspect of an event which makes it unique, cannot, therefore, be the subject of scientific investigation. For an ensemble of unique events *has* no common denominator, and there is nothing in it to explain; such events are *random*, and the observer perceives them as noise. Now, since every real event incorporates *some* element of uniqueness, every ensemble of real events contains some noise. And so the basic problem of scientific investigation is to recognize a significant structure of an ensemble of events above its inevitable background noise. Hence the lower the background noise of a natural phenomenon--that is, the smaller the role of the uniqueness of its constituent events in the overall picture--the more unambiguous is its structure.

Most of the natural phenomena for which successful scientific theories had been worked out prior to about a hundred years ago are relatively noise-free. Such phenomena were explained in terms of *deterministic* laws, which assert that a given set of initial conditions (antecedent situation) can lead to one and only one final state (consequent). But toward the end of the nineteenth century, the methods of mathematical statistics came to be trained on previously inscrutable

---

[73]Mandelbrot, B. (1963). New Methods in Statistical Economics. Journal of Political Economy 71:421.

[74]Mandelbrot, B. (1977). Fractals: Form, Chance, and Dimension. W. H. Freeman and Company, San Francisco.

physical phenomena involving an appreciable element of noise.   This development gave rise to the appearance of *indeterministic* laws of physics, such as the kinetic theory of gases and quantum mechanics. These indeterministic laws envisage that a given set of initial conditions can lead to several alternative final states. An indeterministic law is not devoid of predictive value, however, because to each of the several alternative final states there is assigned a probability of its realization.

Mandelbrot asserts that many of those noisy phenomena which continue to elude successful theoretical analysis will not only be inaccessible to deterministic theories, but will prove refractory also to formulation in terms of ordinary, or 'first-stage,' indeterministic theories. Instead, such noisy phenomena belong to the realm of what Mandelbrot calls 'second-stage' indeterminism. The criterion by which indeterministic phenomena of the first stage are distinguished from those of the second stage is the statistical character of the noise inherent in the ensemble of events, or its *spontaneous activity*. In almost all phenomena for which it *has* so far been possible to make successful first-stage indeterministic scientific theories, the spontaneous activity displays a statistical distribution such that the mean value of a series of observations converges rapidly toward a limit. For instance, the spontaneous activity of a gas satisfies this condition: here the kinetic energy of individual gas molecules is subject to a very wide variation: nevertheless the mean energy per molecule converges to a limit. But many of the phenomena for which it has *not* been possible make successful scientific theories so far turn out to possess a spontaneous activity which displays quite a different distribution. For such phenomena, the mean value of a series of observations converges only very slowly, or not at all, toward a limit. In the absence of such a limit, according to Mandelbrot, it is very difficult to ascertain whether any structure the observer believes he has perceived is real or merely a figment of his imagination.

A statistical distribution of this kind is called "Pareto" distribution after the turn-of-the-century Italian economist who first observed it in the distribution of incomes. As it turns out, many geophysical, meteorological, and astrophysical phenomena, such as size of mineral deposits, annual rainfall, and energies of meteorites and cosmic rays, follow Pareto distributions. In the case of such phenomena, the number of observations available to the scientist for analysis is usually far too small to permit meaningful estimates of the mean, or typical values of significant parameters. Accordingly, the perception of structure in these phenomena is very difficult; there is no guarantee that they are *not* due to pure chance.

Second stage indeterminism accounts for the conspicuous lack of successful theories in the social sciences, such as economics and sociology, as compared to the much more successful natural sciences. The reason for this debility of the social sciences is that the interesting phenomena of which the social sciences seek to give a quantitative account are mainly shrouded in the fog of Pareto distributions. For that reason, it may be in vain to expect a flowering of the social sciences, even though we are in such dire need of their answers to 'why?' and 'how?' questions. Most of their theories will necessarily belong to the realm of second-stage indeterminism, beyond the reach of practical validation.

To summarize: *the causally connected events of many phenomena for which we still lack satisfactory theoretical treatments, especially those addressed by the social sciences, are hidden from view in a fog of random noise. The number of observations that are available for the analysis of such noisy phenomena is usually far too small to permit a clear distinction between their random and their causally significant aspects.* *Consequently, scientific theories advanced to account for such phenomena can command assent to only a very limited extent, that is to say, can have little claim to intersubjective, let alone, objective, truth.* Thus we encounter here another limit to indefinite scientific progress. This limit can be designated as a *structural* one, inasmuch as it arises from the difficulty in perceiving the real structure of complex phenomena in the realm of second-stage indeterminism.

6.   HERMENEUTICS

After having discussed a semantic and a structural limit of science, I finally consider a third cognitive limit. That limit has long been known to scholars active in the discipline known as 'hermeneutics.'[75] This designation was originally given by theologians to the theory of interpretation of sacred texts, especially of the Bible. It takes its name from Hermes, the divine messenger. In his capacity as an information channel linking gods and men, Hermes must 'interpret,' or make explicit in terms that ordinary mortal can understand, the implicit meaning that is hidden in the gods' messages. Probably the most important philosophical contribution of hermeneutics is its insight that hidden meanings pose a procedural difficulty for textual interpretation, because one must understand the context in which the whole text is embedded before one can uncover hidden meanings in any of its parts. Here we face a logical dilemma, a vicious hermeneutic circle. On the one hand, the words and sentences of which a text is composed have no meaning until one knows the meaning of the text as a whole. On the other hand, one only can come to know the meaning of the whole text through understanding its parts. To break this vicious circle, hermeneutics invokes the doctrine of *pre-understanding*.

Pre-understanding represents the experience and insights that the interpreter must bring to the task of interpreting a particular text, which permit him to grasp intuitively the sense of the text as a whole. In assessing the epistemological status of hermeneutic studies, we may ask to what extent the concept of objective truth is applicable. An objectively true interpretation would presumably be one that has made explicit the 'true' meaning hidden in the text, that is, the meaning intended by the author. But here we encounter two difficulties.

First, the author may not have been--in fact, according to the teachings of analytical psychology, most likely was not--consciously aware of the (subconsciously) intended meaning of his own text. Therefore the outcome of the only operational test of the validity of an interpretation, namely, asking the author: "Is this your intended meaning?" or discovering the author's own explicit statement of the meaning of his text, does not provide an objective criterion of interpretative truth. What would be needed in addition is an (also interpretative) exploration of the author's subconscious.

Second, in order to be eligible for attempting a true interpretation in the first place, the interpreter must possess just those experiences and insights that the

---

[75]Gadamer, H. G. (1976). Philosophical Hermeneutics (D. E. Linge, transl. and ed.), University of California Press, Berkeley, 243 pp.

author presupposed (consciously or subconsciously) in the audience to which his text is addressed. But those experiences and insights, that is, the interpreter's pre-understanding, are necessarily based on his own subjective historical, social, and personal background. Hence an interpretation could command assent only among person who happened to bring the same pre-understanding to the text.

Because of the conceptual lack of an operational test for truth, on the one hand, and the necessarily subjective nature of pre-understanding on the other there cannot be such a thing as an objectively true interpretation. Contemporary students of hermeneutics, such as Hans-Georg Gadamer,[76] assert that the very concept of the true meaning is incoherent, inasmuch as they claim that a text can have very many meanings, of which the author's intended meaning is only one, and not even a privileged one. It is this evident unattainability of objective truth in interpretation that seems to make hermeneutics different from the Greek conception of science, for which the belief in the attainability of objectively true propositions about the world is metaphysical bedrock.

To what extent is this traditional belief actually justified? According to some contemporary philosophers of science, such as Thomas Kuhn[77] and Paul Feyerabend,[78] it is not justified, because the scientist also must bring subjective notions equivalent to hermeneutic pre-understanding to his search for explanations of phenomena. Indeed, as was pointed out in the early 1930s by Ludwik Flek[79] (the still unappreciated precursor of these latter-day radical critics of eternal truth of scientific explanation) even the so-called facts of science are not objective givens, but socially conditioned creations.

Even if we admit this radical critique of the Greek conception of science, the fact remains nevertheless that some scientific propositions are based on less pre-understanding than others, and hence can lay claim to a relatively greater approach to objective truth. We can estimate the degree to which a scientific proposition might be objectively true by assessing the extent to which pre-understanding played a role in its development. Such an assessment shows why the Greek belief in the possibility of objectively true scientific propositions is somewhat more justified in the 'hard' natural sciences, such as physics, than in the 'soft' human and social sciences, such as psychology, economics and sociology. One of the main reasons for these differences in the degree to which claim can be laid to objective truth is that the phenomena which the 'soft' sciences seek to explain are much more complex than those addressed by the 'hard' sciences.[80]

---

[76]Gadamer, H. G. (1976. Philosophical Hermeneutics (D. E. Linge, trans. and ed.), University of California Press, Berkeley, 243 pp.

[77]Kuhn, T. S. (1964). The Structure of Scientific Revolutions. University of Chicago Press.

[78]Feyerabend, P. (1975). Against Method. Atlantic Highlands, Humanities Press, N.J.

[79]Fleck, L. (1979). Genesis and Development of a Scientific Fact (T. J. Trenn and R. K. Merton, eds.; F. Bradley and T. J. Trenn, trans.), University of Chicago Press, Chicago, 203 pp.

[80]Stent, G. S. (1984). Hermeneutics and the analysis of complex biological systems. In: Evolution at a Crossroads: The New Biology and the New Philosophy of Science (D. H. Depew

By way of comparing a pair of extreme examples--one very hard, the other very soft--we may consider mechanics and psychoanalysis. There is an aura of objective truth about the theories of classical mechanics because the phenomena which mechanics consider significant, such as steel balls rolling down inclines, are of low complexity. It is therefore possible to dissect a phenomenon into the essential components--steel ball and incline--that are governed by the causal connections envisaged by theory, without having to invoke much pre-understanding. To validate the theory, one can adduce critical observations and experiments with various kinds of steel balls and inclines. By contrast, there is no comparable aura of truth about the propositions of analytical psychology, because the phenomena of the psyche which it attends are very complex. Without resort to far-reaching pre-understanding, the analyst cannot recognize any structures in, let alone dissect, the phenomenon that is his analysis and into its essential, causally connected components. In psychoanalysis, there are no critical observations or experiments, because the failure of any prediction based on psychoanalytic theory can almost always be explained away by modifying slightly one's pre-understanding of the phenomenon. Hence psychoanalytical theories are mainly beyond the reach of validation, which is why many scientists deny--in my opinion unjustifiably--psychoanalysis' standing as a scientific discipline.

Neurobiology, the branch of science in which I myself happen to be working, spans a broad range on this hardness-softness scale of the sciences. At its hard end, neurobiology is represented by electrophysiological, anatomical and biochemical studies of nerve cells. Although the phenomena associated with nerve cells are more complex than rolling steel balls, they can still be accounted for in terms of theories that are susceptible to seemingly objective proof. But at its soft and more fascinating end, neurobiology is represented by system-analytical studies of the structure and function of large and very complicated cellular networks. And the phenomena associated with neural networks approach the psyche in their complexity; in fact, they *include* the psyche. Hence at its soft end, neurobiology takes on some of the characteristics of hermeneutics: the student of a complex neural network must bring extensive pre-understanding to the system as a whole before he can attempt to interpret the function of any of its parts. Accordingly, the propositions that are put forward about complex neural systems are likely to lack the appearance of objective truths.[81]

7.   CODA

I conclude my discussion of the cognitive limits relevant to the question "The End of Science?" by recalling once more that science is a semantic activity whose purpose is to communicate truths about the lawful connections between events in the world of things. Science seeks to find answers to theoretical 'why?'

---

and B. H. Weber, eds.). M.I.T. Press/Bradford Books, Cambridge, Massachusetts, pp. 209-225.
Hermeneutics and the analysis of complex biological systems. Proc. Am. Phil. Soc. 130:336-342.

[81]Stent, G. S. (1984). Hermeneutics and the analysis of complex biological systems. In:
Evolution at a Crossroads: The New Biology and the New Philosophy of Science (D. H. Depew
and B. H. Weber, eds.). M.I.T. Press/Bradford Books, Cambridge, Massachusetts, pp. 209-225.
Hermeneutics and the analysis of complex biological systems. Proc. Am. Phil. Soc. 130:336-342.

questions, which may, in their turn, lead to answers to practical 'how?' questions. The increase in power over the world of things afforded by an answer to a 'how?' question provides, without need for tightly reasoned logical inferences, an ontological validation of the truth of the answer to the antecedent 'why?' question. This question and answer process is rooted in the intuitive epistemological concepts of which we take possession in the course of the cognitive development in our childhood. Thanks to that conceptual endowment, the project of science founded by the Greeks went from success to success, so that by the end of the 19th century it had become perfectly clear that not only is nature highly accessible to our intuitive concepts but that scientific knowledge can confer on man very extensive power over the world of things.

In the course of the 20th century, however, as science continued to pursue the secrets of nature to the bottom of the night, it began to run into some cognitive limits. I have discussed three such limits. The first is a semantic limit, which we encounter in dimensional domains that lie far beyond our direct experience. There we fail in our capacity to represent the world in linguistic images that can be brought into harmony with our internalized, intuitive picture of the world. In these remote domains, we cannot attain that which we call 'truth.' The second cognitive limit is a structural one, which devolves from the impenetrable statistical character of many, as yet unaccounted for phenomena. In the study of these phenomena, it is difficult to convince ourselves that a perceived structure is a genuine constituent of the phenomenon and not merely a figment of our imagination. The third cognitive limit is a subjective one, which arises from the unavoidable recourse to subjective pre-understanding in the interpretation of highly complex phenomena.

It would appear, therefore, that *quite apart from any physical, social or political limits that may exist, science is approaching purely cognitive limits in our days. On the one hand, recent scientific progress has made it necessary to alter the intuitive concepts by means of which we come to know the world in the first place and has introduced into scientific discourse a semantically meaningless parlance. On the other hand, doubts have arisen whether the structures really exist that we perceive in the constituent events of complex, as-yet-not-successfully-fathomed phenomena. Science has achieved very much and it should still achieve much more. But what will it all mean?*

QUESTION

HOLTON: I was very intrigued by Professor Stent's lecture. I'm very glad he talked about the limits of science. He gave us a structure for organizing our thoughts about limits which seems to me very useful. I have a problem, however, with one of his examples, that of chromodynamics. The point, as I understand it, is that a gap has occurred between our intuitive perceptions and what is out there. Whereas the electrons could still be vaguely visualized as little green spheres, for quark physics you must speak of up and down, and spin, and colour, which are arbitrary concepts. So Stent says, we have come up against a limit. Well, I have two problems with that. First, in some sense, even the concepts with which we are quite familiar now in science are essentially arbitrary. Einstein had a very nice way of putting it. He said that the way to think about our concepts, even time and space, is not to think of them as having been extracted from experience the way you extract soup bouillon from beef. Rather, the relationship is that of the number on the check that you hand in for your coat. It's an arbitrary relation between the check and the coat. All that counts is that the woman at the checkroom gives you back your coat so you don't freeze. So from the beginning our concepts are abstractions, and it is true that after a while we get used to them. It takes a while to get used to them. Remember Copernicus and his counter-intuitive ideas. Remember Darwin, remember Freud, these counter-intuitive people. Remember relativity, it took ten years for the academies of science to get down to a language in which they did not ask only confused questions about the new categories. Remember Heisenberg, who said the thing we must do is to forget about our visualizable language. We should stick to mathematics.

My second point is that our students are the ones who are at the frontier of intuitive perception. As I see it, they have direct contact with these concepts. They don't see them anymore in the way that some of us saw them when we were young. They see it more directly though the mathematical symmetry of the concepts and to them there is really very little problem. Fifty years from now, there won't be anybody around who worries about it anymore, than we now worry about the motion of the earth.

STENT: No, my belief is that there is really something novel. There is the metaphor and there is the semantic nexus and they are connected in some kind of intuitive way. And my feeling is that in common dynamics, it is really something new that is happening. I'm talking about the predicates. I'm not talking about the name quark, that's arbitrary. But I'm talking about the predicates, the qualities that are assigned to these things. And I think it's new that they have the words which have no nexus whatsoever, which are totally arbitrary. I don't think this is what Einstein could have meant because it has nothing to do with a retrieval system. To think about reality, to think about quarks, my belief is that there has to be some metaphor which connects that with everyday talk. My feeling is that these young physicists we are discussing are more and more like pure mathematicians. I don't consider mathematicians scientists because they are doing something else. My belief is that current dynamics is more like that mathematical activity which is divorced from reality. I believe it is new that predicates are being used to describe components in reality in a metaphorical way where there is no nexus for a meaningful metaphor. That's why I consider this very serious.

HEIDCAMP: This is a question from our audience. It seems to me that present day physicists are almost able to visualize concepts like curvature of space-time, wave-particle duality. In light of this, it may be possible that as science discovers new laws or ideas, our intuitive range will be enhanced accordingly. And in that sense there may not be an ultimate somatic limit but only one relative to present time.

STENT: This is relevant to the struggle we just had here. What the present-day physicists are able to accommodate is the incoherences or the contradictions between the intuitive meaning of these things and what they're supposed to mean in physical theory. But to make this accommodation it has to bring something into your mind. Whereas, I believe that if charm, up and down, etc. bring anything to your mind, it's wrong, because it has nothing to do with those things.

Contributors

Sheldon Lee Glashow is Higgins Professor of Physics and Mellon Professor of the Sciences at Harvard University. He is also an affiliated senior scientist at the University of Houston and a distinguished visiting scientist at Boston University. In 1979 he, along with Abdus Salam and Steven Weinberg, was awarded the Nobel Prize in Physics for research on elementary particles. Dr. Glashow is honored as a fellow in the American Association for the Advancement of Science and the American Physical Society, and is a member of the National Academy of Sciences, the American Academy of Arts and Sciences and Sigma Xi. He is author or co-author of over 200 journal articles and book chapters, and in 1988 published <u>Interactions: A Journey Through the Mind of a Particle Physicist and the Matter of His World</u>.

Ian Hacking is a professor in the Institute for the History and Philosophy of Science and Technology and the department of philosophy at the University of Toronto. In 1986 he was named a fellow of the Royal Society of Canada. Before taking an appointment at Toronto, Dr. Hacking was a professor of philosophy at Stanford University. He has written over 100 book chapters, journal articles and reviews, and has edited <u>Scientific Revolutions</u> (1981) and <u>Exercises in Analysis by Students of Casimir Lewy</u> (1985). Books by Dr. Hacking include <u>Logic of Statistical Inference</u> (1965), <u>A Concise Introduction to Logic</u> (1972), <u>Why Does Language Matter to Philosophy?</u> (1975), <u>The Emergence of Probability</u> (1975) and <u>Representing and Intervening</u> (1983).

Sandra Harding is professor of philosophy and director of women's studies at the University of Delaware. She serves as a member of the editorial boards of a number of journals, including <u>Feminist Studies</u>, <u>Hypatia: A Feminist Journal of Philosophy</u> and <u>The Women's Review of Books</u>. Dr. Harding is the author of <u>The Science Question in Feminism</u> (1986), which won the Jessie Bernard Award of the American Sociological Association in 1987 and was named one of the five best books of 1986 by the <u>Socialist Review</u>. In addition she is editor or co-editor of books including <u>The Human Sciences in Human Perspective</u> (with Marlynn May, 1977), <u>Discovering Reality: Feminist Perspectives on Epistemology, Metaphysics, Methodology and Philosophy of Science</u> (with Marrill Hintikka, 1983) and <u>Feminism and Methodology: Social Science Issues</u> (1987).

Mary Hesse is professor emerita of philosophy of science at Cambridge University and a fellow of the British Academy. She serves on the editorial boards of <u>Studies in the History and Philosophy of Science</u> and <u>Philosophy of Science</u>. Dr. Hesse has written over 100 journal and encyclopedia articles and book chapters, and is the author of books including <u>Science and the Human Imagination</u> (1954), <u>Models and Analogies in Science</u> (1963), <u>The Structure of Scientific Inference</u> (1974), <u>Revolutions and Reconstructions in the Philosophy of Science</u> (1980) and <u>The Construction of Reality: Gifford Lectures at the University of Edinburgh</u> (with M. A. Arbib, 1987). She has been a visiting professor at several American universities, including Yale, Minnesota, Chicago and Notre Dame.

Gerald Holton is Mallinckrodt Professor of Physics and professor of history of science at Harvard University. He is a fellow of the American Academy of Arts and Sciences, the American Association for the Advancement of Science and the American Physical Society, among others. Dr. Holton has earned professional honors including the Robert A. Millikan Medal (1967), a Presidential Citation for Service to Education (1984), the John P. McGovern Medal of Sigma Xi (1985) and the Andrew Gemant Award from the American Institute of Physics (1989). Professional journals on whose editorial boards Dr. Holton serves include Science, Technology and Human Values, Daedalus and Minerva. His publications include Thematic Origins of Scientific Thought: Kepler to Einstein (1973, 1988), The Scientific Imagination: Case Studies (1978), Albert Einstein: Historical and Cultural Perspectives (1982) and The Advancement of Science and Its Burdens: The Jefferson Lecture and Other Essays (1986).

Gunther S. Stent is professor of molecular biology at the University of California, Berkeley. He serves as chair of the neurobiology section of the National Academy of Sciences and is an external member of the Max Planck Institute for Molecular Genetics in Berlin. Dr. Stent is honored as a fellow of the Institute for Advanced Study in Berlin and the American Society for the Advancement of Science. He is a member of learned societies including the National Academy of Sciences, the American Philosophical Society and the Society for Neuroscience, and he has served as a member of the Basic Research Advisory Committee for the March of Dimes since 1975. Dr. Stent is editor of Function and Formation of Neural Systems (1977), A Critical Edition of J. D. Watson's "The Double Helix" (1980) and Max Delbruck's "Mind from Matter" (1986), and author of books including The Coming of the Golden Age (1969) and Paradoxes of Progress (1978).

Janine Marie Genelin, St. Peter, secretarial center supervisor, whose genuine goodwill and consummate computer skills, prepared this manuscript for publication.